ADVANCES IN SOLITON RESEARCH

ADVANCES IN SOLITON RESEARCH

L.V. CHEN
EDITOR

Nova Science Publishers, Inc.
New York

For permission to use material from this book please contact us:
Telephone 631-231-7269; Fax 631-231-8175
Web Site: http://www.novapublishers.com

LIBRARY OF CONGRESS CATALOGING-IN-PUBLICATION DATA
Advances in soliton research / L.V. Chen (editor).
p. cm.
Includes index.
ISBN 1-59454-769-6
1. Solitons--Research. I. Chen, L. V.
QC174.26.W28A28 2004
531'.1133--dc22 2005027785

Published by Nova Science Publishers, Inc. ✦ New York

CONTENTS

PREFACE

Since their discovery a mere thirty years ago, solitons have been invoked to explain such diverse phenomena as:

The long lived 'giant red spot' in the highly turbulent Jovian atmosphere. The famous Fermi-Pasta-Ulam paradox wherein a nonlinearly coupled lattice of particles does not display the "expected" equipartition of energy among available modes. Ion-acoustic waves in a plasma. Energy storage and transfer in proteins via the Davydov soliton.

The propagation of short laser pulses in optical fibers over long distances with negligible shape change. This new volume presents important research from around the globe.

It is interesting to note that spinning as well as pulsating string solitons in Anti de Sitter backgrounds were considered already ten years ago, as examples of string dynamics in curved spacetimes. In the last two years, these string solitons have attracted enormous attention in connection with the conjectured duality between super string theory on $AdS_5 \times S^5$ and $N = 4$ SU(N) super Yang-Mills theory in Minkowski space. In Chapter 1 the authors start by reviewing some of the known string solitons in AdS_5, and then extend the two-spin solution to $AdS_5 \times S^5$ by including pulsation in S^5. We go through the world-sheet covariant approach to linearized perturbations around string solitons in curved spacetimes, derive and analyse, analytically and numerically, the equations for perturbations around the pulsating two-spin string soliton in $AdS_5 \times S^5$, the main focus being to reach a conclusion as to whether coupling pulsation to the spinning strings leads to stabilisation or the contrary.

As explained in Chapter 2, the space-uniform amplitude envelope of the Ion Temperature Gradient driven turbulence in Tokamak is unstable to small perturbations and evolves to nonuniform, soliton-like modulated profiles. The induced poloidal asymmetry of the transport fluxes can generate spontaneous poloidal spin-up of the tokamak plasma.

In a mean-field approximation Bose-Einstein condensates are described by the Gross-Pitaevskii equation. In quasi-one-dimensional traps this non-linear Schrödinger equation becomes one dimensional and it has dynamically stable dark soliton (kink) solutions. The classical dark soliton propagates across the condensate without any dispersion. However, the Gross-Pitaevskii equation is a mean-field approximation which neglects any depletion of atoms from the condensate wave function. In Chapter 3 the authors show that in the case of dark soliton atomic collisions can lead to substantial depletion from the condensate to the notch of the dark soliton wave function. In the Bogoliubov vacuum state the density of depleted atoms in the soliton notch is, for typical experimental parameters, as large as 14% of the condensate density. This minimal stationary quantum depletion indicates that depletion effects are strong. Indeed, we find that in a dynamical process of experimental generation of the dark soliton the atoms depleted from the condensate to the soliton notch fill up the notch after just 10 milliseconds. This time is in a striking coincidence with the time when the dark solitons are experimentally observed to loose contrast. This prediction is based on an expectation value of the density operator. The expectation value is an average over many experimental realizations. To see what is a generic outcome of a single realization of density measurement we derive *N*-particle wave functions in particle representation and then simulate density measurements using *N*-particle probability distributions that follow from the *N*-particle wave functions. We find that in the case of highly idealized states, like the Bogoliubov vacuum, a single realization of density measurement will always find an empty dark soliton with a random position that fluctuates from experiment to experiment. However, in a generic non-stationary state a single realization of the density measurement will find a dark soliton with a random position but also with a random amount of depleted atoms inside the soliton notch. Pictures of the dark soliton in a generic non-stationary state are loosing contrast.

Fullerenes and nanotubes consist of a large number of carbon atoms sitting on the sites of a regular lattice. For pratical reasons it is often useful to approximate the equations on this lattice in terms of the continuous equation. At the moment, the best candidate for such an equation is the modified non-linear Schrödinger equation. In Chapter 4, the authors study the modified non-linear Schrödinger equation, which arises as continuous equation in a system describing an excitation on a hexagonal lattice. This latter system can e.g. describe electron-phonon interaction on fullerene related structures such as the buckminster fullerene and nanotubes. Solutions of this modified non-linear Schrödinger equation, which have solitonic character, can be constructed using an Ansatz for the wave function, which has time and azimuthal angle dependence introduced previously in the study of spinning boson stars and Q-balls. We study these solutions on a sphere, a disc and on a cylinder.

Our construction suggests that non-spinning as well as spinning solutions, which have a wave function with an arbitrary number of nodes, exist. We find that this property is closely related to the series of well-known mathematical functions, namely the Legendre functions when studying the equation on a sphere, the Bessel functions when studying the equation on a disc and the trigonometric functions, respectively, when studying the equation on a cylinder.

Wrap *m* D5-branes around the 2-cycle of a conifold, place *n* D3-branes at a point and watch the system relax. The D5-branes source *m* units of RR 3-form flux on the 3-cycle, which cause dielectric NS5-branes to nucleate and repeatedly sweep out the 3-cycle, each time gaining *m* units of D3-charge while the stack of D5-branes loses *m* units of D3-charge. A similar description of the Klebanov-Strassler cascade has been proposed by Kachru et al. when $m>>m-n$, where it is a tunneling event in the dual field theory. Using the T-dual MQCD, the authors of Chapter 5 argue that the above process occurs for any *m* and *n* and in particular may continue for more than one step. The nonbaryonic roots of the SQCD vacua lead to new cascades because, for example, the 3-cycle swept does not link all of the D5's. This decay is the S-dual of a MMS instanton, which is the decay into flux of a brane that is trivial in twisted K-theory. This provides the first evidence for the S-dual of the K-theory classification that does not itself rely upon any strong/weak duality.

In: Advances in Soliton Research
Editor: L.V. Chen, pp. 1-43
ISBN 1-59454-769-6

Chapter 1

STABILITY OF PULSATING TWO-SPIN STRING SOLITONS IN $AdS_5 \times S^5$

A.L. Larsen and A. Khan
Physics Department, University of Southern Denmark,
Campusvej 55, 5230 Odense M, Denmark

Abstract

It is interesting to note that spinning as well as pulsating string solitons in Anti de Sitter backgrounds were considered already ten years ago, as examples of string dynamics in curved spacetimes. In the last two years, these string solitons have attracted enormous attention in connection with the conjectured duality between super string theory on $AdS_5 \times S^5$ and $\mathcal{N} = 4$ SU(N) super Yang-Mills theory in Minkowski space. We start by reviewing some of the known string solitons in AdS_5, and then extend the two-spin solution to $AdS_5 \times S^5$ by including pulsation in S^5. We go through the world-sheet covariant approach to linearized perturbations around string solitons in curved spacetimes, derive and analyse, analytically and numerically, the equations for perturbations around the pulsating two-spin string soliton in $AdS_5 \times S^5$, the main focus being to reach a conclusion as to whether coupling pulsation to the spinning strings leads to stabilisation or the contrary.

1 Introduction and Results

In connection with the conjectured duality [1, 2, 3] between super string theory on $AdS_5 \times S^5$ and $\mathcal{N} = 4$ SU(N) super Yang-Mills theory in Minkowski space, there has recently been a lot of interest in string solitons in AdS_5, S^5 and $AdS_5 \times S^5$. Rigidly rotating strings in AdS [4], as well as circular pulsating strings [5] were considered 10 years ago, but the direct connection with gauge theory was only discovered quite recently [6]. Namely, it was noticed that the subleading term in the scaling relation between energy E and spin S, for the rigidly rotating strings, is logarithmic

$$E - S \sim \ln(S) \tag{1.1}$$

similarly to results found for certain Yang-Mills operators back in the early days of QCD [7, 8, 9, 10, 11]. The paper [6] initiated a whole industri of finding new string solitons, and in some cases also the corresponding Yang-Mills operators. See for instance [12, 13, 14, 15, 16, 17, 18, 19, 20, 21, 22, 23, 24, 25, 26] for various string configurations.

More recently it was discovered that multi spin solutions could also be constructed easily [27]. Among others, a relatively simple solution was found in AdS_5 describing a string which is located at a point in the radial direction, winding around an angular direction and spinning with equal angular momentum in two independent planes. In the long string limit it was shown that

$$E - 2S \sim S^{1/3} \tag{1.2}$$

differing substantially from (1.1). This is not at all a problem though, since the subleading terms are well-known to depend on the particular string configuration. Multi spin solitons have been further investigated in a number of papers including [26, 28, 29, 30, 31, 32, 33, 34, 35, 36] (for a review see [37]). In particular, in [32] it was shown that the solutions to the string equations of motion can be classified and related to solutions of the integrable Neumann system.

In the present paper we first present most of the known string solitons in AdS_5. In particular, we construct another one-parameter family of multi spin solutions in AdS_5. These solutions describe $2N$-folded strings in the radial direction and also winding M times around an angular direction. They are spinning with equal angular momentum in two independent planes. Moreover, these strings are quite different from the usual straight strings spinning in AdS [4], since they are not spinning around their center. Instead, the whole string is spinning, or actually orbiting, around origo similarly to solutions previously found in black hole spacetimes [4, 15, 38, 39].

For long strings we find that

$$E - 2S \sim \ln(S) \tag{1.3}$$

similarly to (1.1) except for the obvious factor 2, but completely different from (1.2). Thus we recover the remarkable logaritmic scaling relation from the one-spin case [6].

The string solitons in AdS_5 can be extended to string solitons in $AdS_5 \times S^5$ in various ways, for instance by adding pulsation or up to 3 R-charges in S^5. Here we are particularly interested in finding out whether pulsation improves the stability properties of string solitons in AdS_5. As an example, we therefore first extend the simplest two-spin string soliton in AdS_5 [27] to include pulsation in S^5, for which it is an open problem whether the pulsation improves the stability properties or the contrary.

We then go through the world-sheet covariant approach to linearized perturbations around string solitons in curved spacetimes, derive and analyse, analytically and numerically, the equations for perturbations around the pulsating two-spin string soliton in $AdS_5 \times S^5$, the main focus being to reach a conclusion as to whether coupling pulsation to the spinning strings leads to stabilisation or the contrary.

After a tremendous amount of analysis, we finally show that the pulsation indeed improves the stability in the sense that the more pulsation we have, the higher spin we can allow and still have stability. More precisely, we find that the maximal allowed spin increases more or less linearly with the amount of pulsation.

2 String Solitons in AdS_5

In this section we review some of the known string solitons in AdS_5. Let us take an AdS_5 with characteristic length H^{-1}. Then we have the line-element

$$ds^2 = -(1+H^2r^2)dt^2 + \frac{dr^2}{1+H^2r^2} + r^2(d\beta^2 + \sin^2\beta d\phi^2 + \cos\beta d\tilde{\phi}^2) \tag{2.4}$$

The 't Hooft coupling in this notation is $\lambda = (H^2\alpha')^{-2}$, where $(2\pi\alpha')^{-1}$ is the string tension. In the standard parametrisation of S^3 we have the ranges $\beta \in [0, \pi/2]$, $\phi \in [0, 2\pi]$, $\tilde{\phi} \in [0, 2\pi]$, but in some cases we shall use the alternative ranges $\beta \in [0, 2\pi]$, $\phi \in [0, \pi]$, $\tilde{\phi} \in [0, \pi]$.

Before we look at the solitons, we start with some general remarks about strings in curved spacetimes. Using the conformal gauge

$$G_{\mu\nu}\dot{X}^\mu X'^\nu = 0, \quad G_{\mu\nu}(\dot{X}^\mu\dot{X}^\nu + X'^\mu X'^\nu) = 0 \tag{2.5}$$

where dot and prime denote derivatives with respect to the world-sheet coordinates τ and σ, the Polyakov action

$$S = -\frac{1}{4\pi\alpha'}\int d\tau d\sigma \sqrt{-h}h^{\alpha\beta}G_{\mu\nu}X^\mu_{,\alpha}X^\nu_{,\beta} \tag{2.6}$$

reduces to

$$S = \frac{1}{4\pi\alpha'}\int d\tau d\sigma G_{\mu\nu}(\dot{X}^\mu\dot{X}^\nu - X'^\mu X'^\nu) \tag{2.7}$$

The equations of motion are obtained using variation. One finds

$$\delta S = \frac{1}{2\pi\alpha'}\int d\tau d\sigma \left[G_{\mu\nu}(\dot{X}^\mu\delta\dot{X}^\nu - X'^\mu\delta X'^\nu) + \frac{1}{2}G_{\mu\nu,\rho}\delta X^\rho(\dot{X}^\mu\dot{X}^\nu - X'^\mu X'^\nu)\right] \tag{2.8}$$

The partial integration produces extra terms since $G_{\mu\nu}$ is not a constant. Using that $\dot{G}_{\mu\nu} = G_{\mu\nu,\rho}\dot{X}^\rho$ etc, one finds the equations of motion

$$\ddot{X}^\mu - X''^\mu + \Gamma^\mu_{\rho\sigma}(\dot{X}^\rho\dot{X}^\sigma - X'^\rho X'^\sigma) = 0 \tag{2.9}$$

where we skipped the surface terms since we consider closed strings. It is straightforward to write down these equations for AdS_5, but let us return to the action with arbitrary metric $G_{\mu\nu}$

$$S = -\frac{1}{4\pi\alpha'}\int d\tau d\sigma \eta^{\alpha\beta}G_{\mu\nu}X^\mu_{,\alpha}X^\nu_{,\beta} \tag{2.10}$$

where $\eta_{\alpha\beta} = diag(-1, 1)$. Now assume that the metric does not depend on the coordinate X^k. Then the action is invariant under the global transformation

$$X^k \longrightarrow X^k + \epsilon \tag{2.11}$$

where ϵ is a constant. We then have a conserved Noether current given by

$$J_k^\alpha = -\frac{1}{2\pi\alpha'}\eta^{\alpha\beta}G_{\mu k}X^\mu_{,\beta} \tag{2.12}$$

that is: $\partial_\alpha J_k^\alpha = 0$. The corresponding constant charge is given by

$$Q_k = \int_0^{2\pi} d\sigma J_k^\tau \tag{2.13}$$

As an example, we take $2+1$ dimensional Minkowski space in polar coordinates. Then we have that the metric only depends on r. It means that we have invariance under constant translations of t and ϕ, with corresponding constant charges Q_t and Q_ϕ. These charges are identified by (minus) the energy E and the angular momentum S (which we call spin), respectively. The computation of energy and spin will be essential in the following subsections.

2.1 The Circular String

We shall now look at some special solutions to these equations. The first example is a circular pulsating string, which is obtained using the ansatz

$$t = t(\tau), \quad r = r(\tau), \quad \beta = \pi/2, \quad \phi = \sigma, \quad \tilde{\phi} = const. \tag{2.14}$$

For this string, using (2.5) and (2.9) we end up with the following equations (the Christoffel symbols etc for AdS_5 are given in the appendix)

$$\frac{\dot{r}^2}{1+H^2r^2} + r^2 - \frac{c_0^2}{1+H^2r^2} = 0 \tag{2.15}$$

$$\dot{t} = \frac{c_0}{1+H^2r^2} \tag{2.16}$$

where c_0 is an integration constant. To solve equation (2.15), we look for solutions of the form

$$r(\tau) = A\mathrm{cn}(B\tau|m) \tag{2.17}$$

where A and B are constants and m is the elliptic parameter. Then we find

$$A = \frac{1}{\sqrt{2}H}\left((4H^2c_0^2+1)^{1/2} - 1\right)^{1/2} \tag{2.18}$$

$$B = (4H^2c_0^2+1)^{1/4} \tag{2.19}$$

$$m = \frac{1}{2}\left(1 - \sqrt{\frac{1}{4H^2c_0^2+1}}\right) \tag{2.20}$$

Then we can integrate (2.16)

$$\begin{aligned} t &= c_0 \int_0^\tau \frac{ds}{1+H^2A^2 - H^2A^2\mathrm{sn}^2(Bs|m)} \\ &= \frac{c_0}{B(1+H^2A^2)} \int_0^{B\tau} \frac{dw}{1-n\mathrm{sn}^2(w|m)} \\ &= \frac{c_0}{B(1+H^2A^2)} \Pi(n; B\tau|m) \end{aligned} \tag{2.21}$$

where $n = H^2A^2/(1+H^2A^2)$. The circular string oscillates between 0 and r_{max} where

$$r_{max} = A \tag{2.22}$$

A large (small) string means therefore large (small) c_0. It is straightforward to compute the constant charges from (2.13). The energy is $E = c_0/\alpha'$ and there is no spin, $S = 0$.

The circular pulsating string was originally already discussed 10 years ago [5] as an example of strings in curved spacetime. More recently ([6, 17] and many others), it was rediscovered in connection with the AdS/CFT correspondence.

2.2 The Rigidly Rotating String

As another example, consider the rigidly rotating string obtained by the ansatz

$$t = t(\tau), \quad r = r(\sigma), \quad \beta = \pi/2, \quad \phi = \omega\tau, \quad \tilde{\phi} = const. \tag{2.23}$$

where ω is a constant. For the rigidly rotating string, using (2.5) and (2.9) we end up with the equation

$$-(1+H^2r^2)c_0^2 + \frac{r'^2}{1+H^2r^2} + r^2\omega^2 = 0 \tag{2.24}$$

where c_0 is again an integration constant, and we then look for solutions of the form

$$r(\sigma) = A\mathrm{cn}(B\sigma|m) \tag{2.25}$$

Then we find

$$\begin{aligned} A &= \left(\frac{c_0^2}{\omega^2 - H^2c_0^2}\right)^{1/2} && (2.26) \\ B &= \omega && (2.27) \\ m &= \frac{H^2c_0^2}{\omega^2} && (2.28) \end{aligned}$$

However, we get one more condition from the periodicity $r(\sigma) = r(\sigma + 2\pi)$, namely

$$2\pi B = 4K(m) \tag{2.29}$$

This condition fixes, at least in principle, ω in terms of Hc_0. Notice also that this condition ensures that $0 \leq m \leq 1$ and thus that the solution is real. The string then is stretched from $-r_{max}$ to $+r_{max}$, where

$$r_{max} = A \tag{2.30}$$

A long (short) string means large (small) c_0. From (2.29) then follows that a long string has $\omega \approx Hc_0$, while a short string has $\omega \approx 1$. In other words, a long (short) string corresponds to $m \approx 1$ ($m \approx 0$).

For both circular and rigidly rotating strings, we notice that small/short strings are almost Minkowskian. This was to be expected since small/short strings do not feel the curvature.

Next, we have to compute the energy and angular momentum. It turns out that they can be expressed in terms of complete elliptic integrals of the parameter defined in (2.28). More precisely, (2.13) gives

$$E = \frac{2\sqrt{m}}{\pi\alpha' H(1-m)} E(m) \tag{2.31}$$

$$S = \frac{2}{\pi\alpha' H^2(1-m)} (E(m) - (1-m)K(m)) \tag{2.32}$$

First consider short strings ($m \approx 0$). Then we get $E \approx \sqrt{m}/H\alpha'$ and $S \approx m/2H^2\alpha'$. It follows that $\alpha' E^2 \approx 2S$, which is just the Minkowski result.

Let us now consider the rigidly rotating strings in the long string limit ($m \approx 1$). Thus we expand energy and angular momemtum near $m = 1$. We will do it in some detail since this computation will be typical for the following subsections. We find (prime means derivative with respect to m)

$$\begin{aligned} E &= \frac{2}{\pi\alpha' H(1-m)} \left(1 - \frac{1}{2}(1-m) + ...\right) (1 - E'(1)(1-m) + ...) \\ &= \frac{2}{\pi\alpha' H(1-m)} \left(1 + \frac{1}{4}(1-m) \ln \frac{16}{1-m} + ...\right) \end{aligned} \tag{2.33}$$

$$\begin{aligned} S &= \frac{2}{\pi\alpha' H^2(1-m)} ((1 - E'(1)(1-m) + ...) \\ &- (1-m)(K(1) - (1-m)K'(1) + ...)) \\ &= \frac{2}{\pi\alpha' H^2(1-m)} \left(1 - \frac{1}{4}(1-m) \ln \frac{16}{1-m} + ...\right) \end{aligned} \tag{2.34}$$

where we used that

$$\begin{aligned} K'(m) &= \frac{-1}{2m} \int_0^1 dx (1 - mx^2 - 1) \frac{1}{\sqrt{(1-mx^2)^3(1-x^2)}} \\ &= \frac{-1}{2m} (K(m) - \Pi(m, m)) \end{aligned} \tag{2.35}$$

$$E'(m) = \frac{1}{2m}\int_0^1 dx(1 - mx^2 - 1)\frac{1}{\sqrt{(1-mx^2)(1-x^2)}} = \frac{1}{2m}(E(m) - K(m)) \quad (2.36)$$

and we only kept the leading and the dominant subleading term. For instance, we kept $(1-m)\ln(1-m)$ but we skipped $(1-m)$.

It follows that

$$E/H - S = \frac{1}{\pi\alpha' H^2}\ln\frac{16}{1-m} + ... = \frac{1}{\pi\alpha' H^2}\ln(\pi\alpha' H^2 S) + ... \quad (2.37)$$

where the last equality follows from using the dominant contribution in (2.34) and by skipping a constant (the factor in front of S can be changed by adding a constant).

The rigidly rotating string was also already discussed quite some time ago [4, 38]. And it was also rediscovered in connection with the AdS/CFT correspondence [6], where in particular the logaritmic behaviour (2.37) was noticed.

2.3 Two-Spin Solutions

It is also possible to construct string solitons with two spins. In the simplest case, the ansatz is

$$t = c_0\tau, \quad r = r_0 = const., \quad \beta = \sigma, \quad \phi = \omega\tau, \quad \tilde{\phi} = \omega\tau, \quad (2.38)$$

This is a circular string in AdS_5 spinning in two different directions.

The r equation becomes

$$\omega^2 = 1 + H^2 c_0^2 \quad (2.39)$$

while the non-trivial conformal gauge constraint is

$$c_0^2 = 2r_0^2 \quad (2.40)$$

The energy E and the 2 spins $S_1 = S_2 \equiv S$ are easily computed

$$E = \frac{(1 + H^2 r_0^2)\sqrt{2H^2 r_0^2}}{H\alpha'} \quad (2.41)$$

$$S = \frac{r_0^2}{2\alpha'}\sqrt{2r_0^2 H^2 + 1} \quad (2.42)$$

For short strings (say $Hr_0 << 1$) we get

$$S \approx \frac{r_0^2}{2\alpha'}(1 + H^2 r_0^2) \quad (2.43)$$

such that

$$H^2 r_0^2 \approx 2H^2\alpha' S - 4H^4\alpha'^2 S^2 \tag{2.44}$$

which inserted into E gives

$$E(S) \approx \frac{1}{H\alpha'}\left(1 + 2H^2\alpha' S - 4H^4\alpha'^2 S^2\right)\sqrt{4H^2\alpha' S - 8H^4\alpha'^2 S^2} \tag{2.45}$$

i.e.

$$E \approx \frac{2\sqrt{S}}{\sqrt{\alpha'}}(1 + H^2\alpha' S) \tag{2.46}$$

which to leading order is just the Minkowski result $\alpha' E^2 = 2(2S)$.

For long strings (say $Hr_0 >> 1$) we get

$$E/H - 2S \approx \frac{\sqrt{2H^2 r_0^2}}{H^2\alpha'} - \frac{H^2 r_0^2}{2H^2\alpha'\sqrt{2H^2 r_0^2}} \tag{2.47}$$

From (2.42) we get

$$S \approx \frac{Hr_0^3}{\sqrt{2}\alpha'}\left(1 + \frac{1}{4H^2 r_0^2}\right) \tag{2.48}$$

such that

$$Hr_0 \approx (\sqrt{2}H^2\alpha' S)^{1/3} - \frac{1}{12(\sqrt{2}H^2\alpha' S)^{1/3}} \tag{2.49}$$

which when inserted into eq.(2.47) gives

$$E/H - 2S \approx \frac{3(\sqrt{2}H^2\alpha' S)^{1/3}}{2^{3/2}H^2\alpha'} - \frac{1}{2^{7/2}H^2\alpha'(\sqrt{2}H^2\alpha' S)^{1/3}} \tag{2.50}$$

This result is valid for $H^2\alpha' S >> 1$.

The multi spin solutions were recently obtained by Tseytlin and Frolov [27, 30, 31]. Thus instead of the logaritmic behaviour (2.37) for long strings, they get a power law. But we will see in the next subsection that this is due to the fact that the string is not extended in the r direction.

It should be noticed that the solution of the present subsection has been generalized to arbitrary unequal angular momenta [34, 37], which gives a somewhat similar (although non-regular) scaling relation.

2.4 More General Two-Spin Solutions in AdS_5

The string which is extended in the r and β directions and spinning with equal spin in the ϕ and $\tilde{\phi}$ directions, is obtained by the ansatz

$$t = c_0\tau, \quad r = r(\sigma), \quad \beta = \beta(\sigma), \quad \phi = \omega\tau, \quad \tilde{\phi} = \omega\tau \tag{2.51}$$

where c_0 and ω are arbitrary constants. Then the r and β equations become

$$r'' - \frac{r'^2 H^2 r}{1+H^2r^2} - \beta'^2(1+H^2r^2)r - H^2c_0^2(1+H^2r^2)r + \omega^2(1+H^2r^2)r = 0 \tag{2.52}$$

$$\beta'' + \frac{2r'\beta'}{r} = 0 \tag{2.53}$$

while the non-trivial conformal gauge constraint is

$$\frac{r'^2}{1+H^2r^2} + r^2\beta'^2 - c_0^2(1+H^2r^2) + r^2\omega^2 = 0 \tag{2.54}$$

The system is reduced to first order form

$$\beta' = \frac{k}{r^2} \tag{2.55}$$

$$\frac{r'^2}{1+H^2r^2} + \frac{k^2}{r^2} + r^2\omega^2 - c_0^2(1+H^2r^2) = 0 \tag{2.56}$$

where k is an integration constant. These equations are identical to those considered in [27] (in the parametrisation $Hr = \sinh\rho$), but only constant r solutions were found there. Similar equations and their generalisations were also considered in [34, 37], but again only constant r solutions were explicitly constructed. We shall now show that solutions with $r = r(\sigma)$ exist as well, provided that the string is folded at least six times in the radial direction.

First concentrate on the r equation (2.56). In the non-negative dimensionless coordinate $y \equiv H^2r^2$, it can be written

$$y'^2 = 4(H^2c_0^2 - \omega^2)y^3 + 4(2H^2c_0^2 - \omega^2)y^2 + 4(H^2c_0^2 - H^4k^2)y - 4H^4k^2 \tag{2.57}$$

If $H^2c_0^2 - \omega^2 = 0$, the solution is

$$y = \frac{-(c_0^2 - H^2k^2)}{2c_0^2} \pm \frac{(c_0^2 + H^2k^2)}{2c_0^2}\cosh(2Hc_0\sigma) \tag{2.58}$$

but it must be discarded since it is not periodic. If $H^2c_0^2 - \omega^2 \neq 0$, we notice that the right hand side can be factorized

$$\begin{aligned} y'^2 &= 4\left(H^2c_0^2 - \omega^2\right)(y+1) \\ &\times \left(y - \left(\frac{H^2(-c_0^2 + \sqrt{c_0^4 + 4k^2(H^2c_0^2 - \omega^2)})}{2(H^2c_0^2 - \omega^2)}\right)\right) \\ &\times \left(y - \left(\frac{H^2(-c_0^2 - \sqrt{c_0^4 + 4k^2(H^2c_0^2 - \omega^2)})}{2(H^2c_0^2 - \omega^2)}\right)\right) \end{aligned} \tag{2.59}$$

To get periodic solutions, we are interested in the case with two non-negative roots, so we get the conditions

$$H^2 c_0^2 - \omega^2 < 0 \tag{2.60}$$

$$c_0^4 + 4k^2(H^2 c_0^2 - \omega^2) \geq 0 \tag{2.61}$$

Now define the function P

$$y = \frac{P}{H^2 c_0^2 - \omega^2} - \frac{2H^2 c_0^2 - \omega^2}{3(H^2 c_0^2 - \omega^2)} \tag{2.62}$$

such that

$$P'^2 = 4P^3 - g_2 P - g_3 = 4(P - e_1)(P - e_2)(P - e_3) \tag{2.63}$$

where the roots $(e_1 > e_2 \geq e_3)$ are

$$e_1 = \frac{1}{3}(2\omega^2 - H^2 c_0^2) \tag{2.64}$$

$$e_2 = \frac{1}{6}H^2 c_0^2 - \frac{1}{3}\omega^2 + \frac{1}{2}H^2\sqrt{c_0^4 + 4k^2(H^2 c_0^2 - \omega^2)} \tag{2.65}$$

$$e_3 = \frac{1}{6}H^2 c_0^2 - \frac{1}{3}\omega^2 - \frac{1}{2}H^2\sqrt{c_0^4 + 4k^2(H^2 c_0^2 - \omega^2)} \tag{2.66}$$

and the invariants are $g_2 = 2(e_1^2 + e_2^2 + e_3^2)$ and $g_3 = 4e_1 e_2 e_3$. Notice also that $\Delta \equiv g_2^3 - 27g_3^2 \geq 0$ (using (2.60), (2.61)).

Equation (2.63) is the Weierstrass equation with solution

$$P(\sigma) = \wp(\sigma + a) \tag{2.67}$$

where $\wp$ is the doubly-periodic Weierstrass function and a is a complex integration constant [40]. To get a real non-singular solution, we take a to be half the imaginary period. Then the Weierstrass function reduces to a Jacobi elliptic function [40]. Before writing down the explicit solution, it is convenient to trade the parameters (c_0, ω, k) for

$$b = e_2 - e_3 = H^2\sqrt{c_0^4 + 4k^2(H^2 c_0^2 - \omega^2)} \tag{2.68}$$

$$m = \frac{e_2 - e_3}{e_1 - e_3} = \frac{2H^2\sqrt{c_0^4 + 4k^2(H^2 c_0^2 - \omega^2)}}{2\omega^2 - H^2 c_0^2 + H^2\sqrt{c_0^4 + 4k^2(H^2 c_0^2 - \omega^2)}} \tag{2.69}$$

$$n = \frac{2\sqrt{c_0^4 + 4k^2(H^2 c_0^2 - \omega^2)}}{c_0^2 + \sqrt{c_0^4 + 4k^2(H^2 c_0^2 - \omega^2)}} \tag{2.70}$$

such that by eqs.(2.60)-(2.61)

$$b \geq 0, \quad 0 \leq m \leq n \leq 1 \tag{2.71}$$

In this parametrisation the solution to (2.56) is

$$H^2r^2(\sigma) = \frac{m}{n-m}\left(1 - n\text{sn}^2\left(\sqrt{\frac{b}{m}}\sigma|m\right)\right) \tag{2.72}$$

so that $Hr(\sigma)$ oscillates between the non-negative values Hr_{max} and Hr_{min}, where

$$Hr_{max} = \sqrt{\frac{m}{n-m}}, \quad Hr_{min} = \sqrt{\frac{m(1-n)}{n-m}} \tag{2.73}$$

This means that the whole string is spinning, or actually orbiting, around origo. It is quite similar to straight strings spinning around black holes [4, 15, 38, 39], but unlike one-spin strings in AdS which spin around their center [4].

The solution (2.72) must be supplemented with the closed string periodicity condition $r(\sigma + 2\pi) = r(\sigma)$. For a $2N$-folded closed string (N positive integer), this condition becomes

$$\frac{\sqrt{m}K(m)}{\pi\sqrt{b}} = \frac{1}{N} \tag{2.74}$$

which trivially gives b in terms of m. This condition of course drops out for $n = 0$ (which implies $b = m = 0$), corresponding to strings which are pointlike in the radial direction [27]. Here we shall only consider strings which are extended in the radial direction.

Now return to the β equation (2.55). It is trivially integrated to

$$\begin{aligned}\beta(\sigma) &= \sqrt{\frac{b(n-m)(1-n)}{nm}}\int_0^\sigma \frac{dx}{1 - n\text{sn}^2\left(\sqrt{\frac{b}{m}}x|m\right)} \\ &= \sqrt{\frac{(n-m)(1-n)}{n}}\Pi\left(n; \sqrt{\frac{b}{m}}\sigma|m\right)\end{aligned} \tag{2.75}$$

Since β is an angular coordinate, we impose the quasi-periodicity condition $\beta(\sigma + 2\pi) = 2M\pi + \beta(\sigma)$, where M is another positive integer which plays the role of a winding number. This condition translates into

$$\frac{\sqrt{(1-n)(n-m)}\Pi(n;m)}{\pi\sqrt{n}} = \frac{M}{N} \tag{2.76}$$

where (2.74) was also used. This equation determines n in terms of m, thus the only free parameter (for fixed N, M) now is m.

The main problem when analysing these solutions is that (2.76) is trancendental. Numerical analysis shows that the left hand side is less than 1/2 (it equals 1/2 only for $m = 0$, corresponding to strings which are pointlike at $r = 0$). Thus, to get non-trivial solutions, we need $N > 2M$ which means that the string must be folded at least six times. In that case we get a one-parameter family of solutions with arbitrary extension in the radial direction.

For short and long strings, we can solve (2.76) analytically and confirm the numerical results. First consider short strings. From (2.72) follows that short strings correspond to $m \approx 0, n \approx 0$ (with $m \leq n$). Using that $\Pi(n;m) \approx \pi/2$ in this case, we get from (2.76)

$$\frac{\sqrt{n-m}}{2\sqrt{n}} \approx \frac{M}{N} \tag{2.77}$$

such that

$$\frac{n}{m} \approx \frac{N^2}{N^2-4M^2}, \quad \text{short strings} \tag{2.78}$$

which implies $N > 2M$. For such short strings, in the limit where r is almost a constant, we get

$$Hr \approx \sqrt{\frac{N^2-4M^2}{4M^2}} \tag{2.79}$$

$$\beta \approx M\sigma \tag{2.80}$$

thus the string is located at a fixed finite value of r and simply winding around in the β direction. This is a special case of the solutions found in [27]. To get all the solutions of [27], one should discard the periodicity condition, as discussed after (2.74). But notice that our notion of short means short in the radial extension, which is different from the notion of short used elsewhere.

Now consider long strings, which correspond to $n \approx 1,\ m \approx 1$ (with $m \leq n$). Here we use the approximation

$$\Pi(n;m) \approx \frac{\pi}{2}\sqrt{\frac{1}{(1-n)(n-m)}}\left(1 - \frac{2}{\pi}\sin^{-1}\sqrt{\frac{1-n}{1-m}}\right) \tag{2.81}$$

Then (2.76) gives

$$\frac{M}{N} \approx \frac{1}{2} - \frac{1}{\pi}\sin^{-1}\sqrt{\frac{1-n}{1-m}} \tag{2.82}$$

such that

$$\sqrt{\frac{1-n}{1-m}} \approx \cos\left(\frac{M}{N}\pi\right), \quad \text{long strings} \tag{2.83}$$

which again implies $N > 2M$. This leads to the following expression for Hr_{min}

$$Hr_{min} \approx \text{cotan}\left(\frac{M}{N}\pi\right) \tag{2.84}$$

Thus the long strings extend outwards from this fixed value. We observe that the long strings come closer to origo than the short strings.

The conserved energy is given by

$$E = \frac{c_0}{2\pi\alpha'}\int_0^{2\pi}(1 + H^2r^2(\sigma))d\sigma \tag{2.85}$$

In our parametrisation, this can be expressed in terms of a complete elliptic integral as follows

$$E = \frac{N\sqrt{(2-n)nm}}{\pi H\alpha'(n-m)}E(m) \tag{2.86}$$

Similarly, the conserved spins are

$$S_1 = \frac{\omega}{2\pi\alpha'}\int_0^{2\pi} r^2(\sigma)\cos^2\beta(\sigma)d\sigma, \; S_2 = \frac{\omega}{2\pi\alpha'}\int_0^{2\pi} r^2(\sigma)\sin^2\beta(\sigma)d\sigma \tag{2.87}$$

such that [32]

$$\frac{E}{H^2 c_0} - \frac{S_1}{\omega} - \frac{S_2}{\omega} = \frac{1}{H^2\alpha'} \tag{2.88}$$

In our case, $S_1 = S_2 \equiv S = (S_1+S_2)/2$, which in terms of complete elliptic integrals gives

$$S = \frac{N\sqrt{n+m-nm}}{2\pi H^2\alpha'(n-m)\sqrt{n}}((m-n)K(m)+nE(m)) \tag{2.89}$$

Further simplifications are obtained for short and long strings, respectively. For short strings, using (2.78), we arrive at the following finite expressions for the energy and spin

$$E \approx \frac{N^2\sqrt{N^2-4M^2}}{4\sqrt{2}H\alpha' M^2} \tag{2.90}$$

$$S \approx \frac{(N^2-4M^2)\sqrt{N^2-2M^2}}{8\sqrt{2}\alpha' H^2 M^2} \tag{2.91}$$

But recall that these are just the minimal values within our family of strings. As explained after (2.80), these strings are not short in the usual sense. In fact, we dont have solutions with $E \approx 0$ and $S \approx 0$ within our family of strings.

For the long strings we can eliminate n using (2.83) and get

$$E \approx \frac{N}{\pi H\alpha' \sin^2\left(\frac{M}{N}\pi\right)(1-m)}\left(1 + \frac{1-m}{4}\ln\frac{16}{1-m}\right) \tag{2.92}$$

$$S \approx \frac{N}{2\pi\alpha' H^2 \sin^2\left(\frac{M}{N}\pi\right)(1-m)}\left(1 - \left(2\sin^2\left(\frac{M}{N}\pi\right) - 1\right)\frac{1-m}{4}\ln\frac{16}{1-m}\right) \tag{2.93}$$

which leads to

$$E/H - 2S \approx \frac{N}{2\pi H^2\alpha'}\ln(2\pi H^2\alpha' S) \tag{2.94}$$

up to an unimportant additive constant.

Thus, we recover the remarkable and celebrated logaritmic scaling relation known from the one-spin case [6]. This should be contrasted with the two-spin solutions considered in [27], where it was found that (in our notation)

$$E/H - 2S \approx \frac{3(H^2\alpha' S)^{1/3}}{2^{4/3}H^2\alpha'} \tag{2.95}$$

The multi-spin solutions of this subsection were recently obtained by the present authors [41]. It is worth noticing that especially the folding number N, but also the winding number M, play a crucial role for the existence of these solutions. The string solitons considered in the previous 3 subsections could also have been equipped with either a folding number or a winding number, but in those cases they would only have played a trivial role.

3 Pulsating Multi Spin Solutions

Multi spin and multi R-charge solutions were recently obtained by Tseytlin and Frolov [27, 30, 31]. Such solutions can easily be combined with pulsation. Here we take for simplicity the simplest two-spin solution in AdS_5, which we considered in Subsection 2.3, and couple it with pulsation in S^5. Let us take an S^5 with radius H^{-1}. Then we get altogether for $AdS_5 \times S^5$ the line-element

$$\begin{aligned} ds^2 &= -(1+H^2r^2)dt^2 + \frac{dr^2}{1+H^2r^2} + r^2(d\beta^2 + \sin^2\beta d\phi^2 + \cos^2\beta d\tilde{\phi}^2) \\ &+ \frac{1}{H^2}(d\theta^2 + \sin^2\theta d\psi^2 + \cos^2\theta(d\psi_1^2 + \sin^2\psi_1 d\psi_2^2 + \cos^2\psi_1 d\psi_3^2)) \end{aligned} \tag{3.1}$$

The ansatz is

$$t = c_0\tau, \quad r = r_0 = const., \quad \beta = \sigma, \quad \phi = \omega\tau, \quad \tilde{\phi} = \omega\tau, \quad \theta = \theta(\tau), \quad \psi = \sigma \tag{3.2}$$

where (c_0, r_0, ω) are constants, and with the remaining coordinates being constants. This is a circular string in AdS_5 spinning in two different directions. It is also a circle in S^5, but pulsating there.

The r and θ equations become

$$\omega^2 = 1 + H^2c_0^2 \tag{3.3}$$

$$\ddot{\theta} + \sin\theta\cos\theta = 0 \tag{3.4}$$

while the non-trivial conformal gauge constraint is

$$\dot{\theta}^2 + \sin^2\theta - H^2c_0^2(1+H^2r_0^2) + H^2r_0^2(1+\omega^2) = 0 \tag{3.5}$$

The θ equation and constraint are solved by

$$\sin\theta(\tau) = \begin{cases} A\text{sn}(\tau|A^2), & A \leq 1 \\ \text{sn}(A\tau|1/A^2), & A \geq 1 \end{cases} \tag{3.6}$$

where

$$A = H\sqrt{c_0^2 - 2r_0^2} \tag{3.7}$$

A is the amplitude of $\sin\theta$ for $A < 1$. For $A > 1$ it is the frequency. More precisely, for $A < 1$ the string oscillates on one hemisphere, for $A = 1$ it starts at one of the poles and approaches for $\tau \to \infty$ the equator while for $A > 1$ it oscillates between the poles.

The energy E and the 2 spins $S_1 = S_2 \equiv S$ are easily computed

$$E = \frac{(1 + H^2 r_0^2)\sqrt{2H^2 r_0^2 + A^2}}{H\alpha'} \tag{3.8}$$

$$S = \frac{r_0^2}{2\alpha'}\sqrt{2r_0^2 H^2 + A^2 + 1} \tag{3.9}$$

For short strings (say $Hr_0 << 1$) we get

$$S \approx \frac{r_0^2}{2\alpha'}\sqrt{A^2+1}\left(1 + \frac{H^2 r_0^2}{A^2+1}\right) \tag{3.10}$$

such that

$$H^2 r_0^2 \approx \frac{2H^2\alpha' S}{\sqrt{A^2+1}} - \frac{4H^4\alpha'^2 S^2}{(A^2+1)^2} \tag{3.11}$$

which inserted into E gives

$$E(S,A) \approx \frac{1}{H\alpha'}\left(1 + \frac{2H^2\alpha' S}{\sqrt{A^2+1}} - \frac{4H^4\alpha'^2 S^2}{(A^2+1)^2}\right)\sqrt{\frac{4H^2\alpha' S}{\sqrt{A^2+1}} - \frac{8H^4\alpha'^2 S^2}{(A^2+1)^2} + A^2} \tag{3.12}$$

For $A = 0$ we get

$$E \approx \frac{2\sqrt{S}}{\sqrt{\alpha'}}(1 + H^2\alpha' S) \tag{3.13}$$

which to leading order is just the Minkowski result $\alpha' E^2 = 2(2S)$. For $A = 1$ we get

$$E \approx \frac{1}{H\alpha'}(1 + 2^{3/2} H^2 \alpha' S) \tag{3.14}$$

and for $A >> 1$

$$E \approx \frac{1}{H\alpha'}(A + 2H^2\alpha' S) \tag{3.15}$$

For long strings (say $Hr_0 >> 1$) we get

$$E/H - 2S \approx \frac{\sqrt{2H^2 r_0^2 + A^2}}{H^2\alpha'} - \frac{H^2 r_0^2}{2H^2\alpha'\sqrt{2H^2 r_0^2 + A^2}} \tag{3.16}$$

Now we have to distinguish between different cases. If $Hr_0 >> A$ we get from (3.9)

$$S \approx \frac{Hr_0^3}{\sqrt{2}\alpha'}\left(1+\frac{A^2+1}{4H^2r_0^2}\right) \tag{3.17}$$

such that

$$Hr_0 \approx (\sqrt{2}H^2\alpha' S)^{1/3} - \frac{A^2+1}{12(\sqrt{2}H^2\alpha' S)^{1/3}} \tag{3.18}$$

which when inserted into (3.16) gives

$$E/H - 2S \approx \frac{3(\sqrt{2}H^2\alpha' S)^{1/3}}{2^{3/2}H^2\alpha'} + \frac{4A^2-1}{2^{7/2}H^2\alpha'(\sqrt{2}H^2\alpha' S)^{1/3}} \tag{3.19}$$

This result is valid for $H^2\alpha' S >> \{1,\ A^3\}$, and therefore holds in particular for $A = 0$ and $A = 1$. Notice also that the pulsation only gives a contribution to the non-leading terms, in this limit. On the other hand, if $Hr_0 << A$ we get from (3.9)

$$S \approx \frac{Ar_0^2}{2\alpha'}\left(1+\frac{r_0^2H^2}{A^2}\right) \tag{3.20}$$

such that

$$H^2r_0^2 \approx \frac{2H^2S\alpha'}{A} - \frac{4S^2\alpha'^2H^4}{A^4} \tag{3.21}$$

and insertion into (3.16)) gives

$$E/H - 2S \approx \frac{A}{H^2\alpha'} + \frac{S}{A^2} \tag{3.22}$$

which holds for $1 << H^2\alpha' S << A^3$. Thus, in this limit, the pulsation completely changes the scaling relation.

All our results reduce for $A = 0$ to those obtained in [27] (and in Subsection 2.3), but otherwise they are quite different. It is straightforward also to generalise the multi R-charge solutions of [27] to include pulsation in AdS_5, but we shall not go into the details here.

4 Perturbations

There are many ways to discuss the perturbations around string solitons. One way is to use the Nambu-Goto action

$$S = -\frac{1}{2\pi\alpha'}\int d\tau d\sigma\sqrt{-det(G_{\mu\nu}X^{\mu}_{,\alpha}X^{\nu}_{,\beta})}$$

and make variations δX^{μ}, thus to obtain $\delta^2 S$, immediately fixing two of them so as to take care of the gauge invariance. This is the method most often used in recent papers. The disadvantages are that it is not world-sheet covariant and that the kinetic energy terms for the perturbations usually come out in a very complicated form.

Another way is to take the Polyakov action in conformal gauge; i.e., one makes a variation of (2.5) and two variations of (2.7) (see [42, 43, 44]). This approach is also used in many papers, but the problem is that one can usually not solve the variation of (2.5), and therefore it is very difficult to separate the physical and unphysical perturbations.

Here we will use the Polyakov action (2.6); i.e., make two variations δX^μ and $\delta h_{\alpha\beta}$. We shall consider only perturbations which are normal to the string world-sheet, thus to ensure that we have purely physical perturbations. This approach is in general world-sheet covariant and we will eventually get the kinetic energy terms in a very simple form (see [45, 46, 47, 48]).

We start by considering a string in flat Minkowski space. The generalization to curved spacetime is then straightforward.

4.1 Minkowski Space

Let us consider the Lagrangian corresponding to the Polyakov action in Minkowski space

$$L = \sqrt{-h}h^{\alpha\beta}g_{\alpha\beta}, \quad g_{\alpha\beta} = \eta_{\mu\nu}X^\mu_{,\alpha}X^\nu_{,\beta} \tag{4.1}$$

We choose to work with the Lagrangian in this section. For the closed strings under consideration here, it means that we dont care about total derivatives. We also take spacetime to have arbitrary dimension D.

The first variation can conveniently be written in the following way

$$\delta L = \sqrt{-h}\left[\left(\frac{1}{2}h^{\alpha\beta}h^{\gamma\delta} - h^{\alpha\gamma}h^{\beta\delta}\right)g_{\gamma\delta}\delta h_{\alpha\beta} + 2\eta_{\mu\nu}h^{\alpha\beta}X^\nu_{,\alpha}(\delta X^\mu)_{,\beta}\right] \tag{4.2}$$

Then the equations of motion are

$$\Box^{(h)}X^\nu = 0 \tag{4.3}$$

and

$$2g_{\alpha\beta} - h_{\alpha\beta}h^{\gamma\delta}g_{\gamma\delta} = 0 \tag{4.4}$$

where we introduced the d'Alambertian

$$\Box^{(h)} = \frac{1}{\sqrt{-h}}\partial_\alpha(\sqrt{-h}h^{\alpha\beta}\partial_\beta) \tag{4.5}$$

The equations of motion for small perturbations δX^μ and $\delta h_{\alpha\beta}$ can be obtained by making variations of (4.3) and (4.4), i.e., by inserting

$$h_{\alpha\beta} = h^{(0)}_{\alpha\beta} + \delta h_{\alpha\beta} \tag{4.6}$$

$$X^\nu = X^\nu_{(0)} + \delta X^\nu \tag{4.7}$$

into (4.3)-(4.4), where $X^\nu_{(0)}$ and $h^{(0)}_{\alpha\beta}$ is an exact solution, and keeping only linear terms in δX^ν and $\delta h_{\alpha\beta}$.

Alternatively, and more elegantly, one can make a second variation of L and interpret the result as a Lagrangian for δX^ν and $\delta h_{\alpha\beta}$, and obtain the equations from there. This is the approach we will follow here. The end result is of course the same.

The second variation of L gives by a direct computation

$$\begin{aligned}
\delta^2 L &= \sqrt{-h}\delta h_{\epsilon\zeta}[\frac{1}{2}h^{\epsilon\zeta}(\frac{1}{2}h^{\alpha\beta}h^{\gamma\delta} - h^{\alpha\gamma}h^{\beta\delta}) \\
&- \frac{1}{2}(h^{\alpha\epsilon}h^{\beta\zeta}h^{\gamma\delta} + h^{\alpha\beta}h^{\gamma\epsilon}h^{\delta\zeta}) \\
&+ (h^{\alpha\epsilon}h^{\gamma\zeta}h^{\beta\delta} + h^{\alpha\gamma}h^{\beta\epsilon}h^{\delta\zeta})]g_{\gamma\delta}\delta h_{\alpha\beta} \\
&+ 4\sqrt{-h}\delta h_{\alpha\beta}(\frac{1}{2}h^{\alpha\beta}h^{\gamma\delta} - h^{\alpha\gamma}h^{\beta\delta})\eta_{\mu\nu}X^\nu_{,\gamma}(\delta X^\mu)_{,\delta} \\
&+ 2\sqrt{-h}h^{\alpha\beta}\eta_{\mu\nu}(\delta X^\nu)_{,\alpha}(\delta X^\mu)_{,\beta}
\end{aligned} \tag{4.8}$$

Using (4.4) and after renaming the indices, we finally obtain

$$\begin{aligned}
\delta^2 L &= \frac{-1}{2}\sqrt{-h}\delta h_{\alpha\beta}(\frac{1}{2}h^{\alpha\beta}h^{\gamma\delta} - h^{\alpha\gamma}h^{\beta\delta})h^{\epsilon\zeta}g_{\epsilon\zeta}\delta h_{\gamma\delta} \\
&+ 4\sqrt{-h}\delta h_{\alpha\beta}(\frac{1}{2}h^{\alpha\beta}h^{\gamma\delta} - h^{\alpha\gamma}h^{\beta\delta})\eta_{\mu\nu}X^\nu_{,\gamma}(\delta X^\mu)_{,\delta} \\
&+ 2\sqrt{-h}h^{\alpha\beta}\eta_{\mu\nu}(\delta X^\nu)_{,\alpha}(\delta X^\mu)_{,\beta}
\end{aligned} \tag{4.9}$$

As mentioned after (4.7), we interpret this as the Lagrangian for the perturbations. It means that δX^μ and $\delta h_{\alpha\beta}$ are the dynamical fields now, while X^μ and $h_{\alpha\beta}$ are fixed background fields. They represent the unperturbed solution, which we denoted by a (0) in (4.6)-(4.7). For simplicity, we do not write the (0) anymore.

The equation of motion for $\delta h_{\alpha\beta}$ is (notice that there are no derivatives of this field in $\delta^2 L$)

$$(\frac{1}{2}h^{\alpha\beta}h^{\gamma\delta} - h^{\alpha\gamma}h^{\beta\delta})h^{\epsilon\zeta}g_{\epsilon\zeta}\delta h_{\gamma\delta} = 2(\frac{1}{2}h^{\alpha\beta}h^{\gamma\delta} - h^{\alpha\gamma}h^{\beta\delta})(X^\nu_{,\gamma}(\delta X_\mu)_{,\delta} + X^\nu_{,\delta}(\delta X_\mu)_{,\gamma}) \tag{4.10}$$

while the equation of motion for δX^μ is

$$\partial_\beta[\sqrt{h}(h^{\alpha\beta}(\delta X^\nu)_{,\alpha} + \delta h_{\gamma\delta}(\frac{1}{2}h^{\alpha\beta}h^{\gamma\delta} - h^{\alpha\gamma}h^{\beta\delta})X^\nu_{,\alpha})] = 0 \tag{4.11}$$

An equation involving only δX^μ is obtained by inserting (4.10) into (4.11). Alternatively we can insert (4.10) back into (4.9)

$$\begin{aligned}
\delta^2 L &= 2\frac{\sqrt{-h}}{h^{\epsilon\zeta}g_{\epsilon\zeta}}(\frac{1}{2}h^{\alpha\beta}h^{\gamma\delta} - h^{\alpha\gamma}h^{\beta\delta})\eta_{\mu\nu}\eta_{\rho\sigma}(X^\nu_{,\gamma}(\delta X^\mu)_{,\delta} + X^\nu_{,\delta}(\delta X^\mu)_{,\gamma}) \\
&\times(X^\sigma_{,\alpha}(\delta X^\rho)_{,\beta} + X^\sigma_{,\beta}(\delta X^\rho)_{,\alpha}) \\
&+ 2\sqrt{-h}h^{\alpha\beta}\eta_{\mu\nu}(\delta X^\nu)_{,\alpha}(\delta X^\mu)_{,\beta}
\end{aligned} \tag{4.12}$$

which only involves δX^μ, and derive the equation from there. In both cases one gets

$$\partial_\beta[\sqrt{-h}(h^{\alpha\beta}(\delta X^\nu)_{,\alpha} + \frac{2}{h^{\epsilon\zeta}g_{\epsilon\zeta}}(\frac{1}{2}h^{\alpha\beta}h^{\gamma\delta} - h^{\alpha\gamma}h^{\beta\delta})X^\nu_{,\alpha}\eta_{\rho\sigma}(X^\sigma_{,\gamma}(\delta X^\rho)_{,\delta} + X^\sigma_{,\delta}(\delta X^\rho)_{,\gamma}))] = 0 \tag{4.13}$$

For a moment we leave the perturbations and return to the unperturbed world-sheet. From the point of view of geometry, we have a 2-dimensional surface Σ embedded in a D-dimenional spacetime M. Thus we have a mapping

$$X^\mu(\sigma^\alpha): \quad \Sigma \to M \tag{4.14}$$

where $\sigma^\alpha = (\tau, \sigma)$. Then we have 2 tangent vectors $X^\mu_{,\alpha}$ to the surface. Now introduce $D-2$ normal vectors

$$\eta_{\mu\nu}N^\mu_i N^\nu_j = \delta_{ij}, \quad \eta_{\mu\nu}N^\mu_i X^\nu_{,\alpha} = 0 \tag{4.15}$$

which also fulfil the completeness relation

$$\eta^{\mu\nu} = \frac{2h^{\alpha\beta}}{h^{\gamma\delta}g_{\gamma\delta}}X^\mu_{,\alpha}X^\nu_{,\beta} + \delta^{ij}N^\mu_i N^\nu_j \tag{4.16}$$

Notice that we have 3 types of indices. The spacetime indices $\mu, \nu, ... = 0, 1, ..., D-1$ which are raised and lowered by the spacetime metric $\eta_{\mu\nu}$, the world-sheet indices $\alpha, \beta, \dots = 0, 1$ (τ and σ) which are raised and lowered by the Polyakov metric $h_{\alpha\beta}$ and finally the indices numbering the normal vectors $i, j, ... = 1, 2, ..., D-2$ which are raised and lowered by the Kronecker δ_{ij}.

To describe the shape of the unperturbed world-sheet, we also define the second fundamental form and the normal fundamental form

$$K^i_{\alpha\beta} = N^i_\mu X^\mu_{,\alpha\beta}, \quad \mu^{ij}_\alpha = N^i_\mu N^{j\mu}_{,\alpha} \tag{4.17}$$

Notice that $K^i_{\alpha\beta} = K^i_{\beta\alpha}$ while $\mu^{ij}_\alpha = -\mu^{ji}_\alpha$. The second fundamental form is also called extrinsic curvature and the normal fundamental form is sometimes called torsion. Together they determine how the world-sheet bends and twists in spacetime.

The Weingarten equations are (D_α is the covariant derivative with respect to the Polyakov metric)

$$D_\gamma X^\mu_{,\beta} = N^\mu_i K^i_{\gamma\beta} + f^\delta_{\gamma\beta}X^\mu_{,\delta} \tag{4.18}$$

$$N^{i\mu}_{,\alpha} = \mu^{ji}_\alpha N^\mu_j - \frac{2}{h^{\epsilon\zeta}g_{\epsilon\zeta}}K^{i\gamma}_\alpha X^\mu_{,\gamma} \tag{4.19}$$

where $f^\alpha_{\beta\gamma}$ is given implicitly by

$$D_\gamma g_{\alpha\beta} = f^\delta_{\gamma\beta}g_{\alpha\delta} + f^\delta_{\gamma\alpha}g_{\beta\delta} \tag{4.20}$$

Notice also that the equation of motion (4.3) is equivalent to $h^{\alpha\beta}K^i_{\alpha\beta} = 0$.

Now return to the perturbations again. Since $(X^\mu_{,\alpha},\ N^\mu_i)$ is a basis for spacetime (at least locally), we can write

$$\delta X^\mu = N^\mu_i \phi^i + X^\mu_{,\alpha}\psi^\alpha \tag{4.21}$$

First consider the tangential perturbations $\delta X^\mu = X^\mu_{,\alpha}\psi^\alpha$. Then (4.10) gives

$$(\frac{1}{2}h^{\alpha\beta}h^{\gamma\delta} - h^{\alpha\gamma}h^{\beta\delta})\delta h_{\gamma\delta} = 2(\frac{1}{2}h^{\alpha\beta}h^{\gamma\delta} - h^{\alpha\gamma}h^{\beta\delta})D_{(c}\psi_{d)} \tag{4.22}$$

This equation is solved by

$$\delta h_{\alpha\beta} = \Lambda h_{\alpha\beta} + 2D_{(\alpha}\psi_{\beta)} \tag{4.23}$$

where Λ is an arbitrary function on the world-sheet. Clearly Λ represents the Weyl symmetry while $(\delta X^\mu,\ \delta h_{\alpha\beta}) = (X^\mu_{,\alpha}\psi^\alpha,\ 2D_{(\alpha}\psi_{\beta)})$ are of course just the reparametrizations, which we exclude.

We are then left with the normal perturbations $\delta X^\mu = N^\mu_i\phi^i$. In this case, (4.10) gives

$$(\frac{1}{2}h^{\alpha\beta}h^{\gamma\delta} - h^{\alpha\gamma}h^{\beta\delta})\delta h_{\gamma\delta} = \frac{4}{h^{\epsilon\zeta}g_{\epsilon\zeta}}K^{\alpha\beta}_i\phi^i \tag{4.24}$$

where we also used (4.3) and (4.4).

Similarly one can obtain the equations of motion for the functions ϕ^i from (4.13). However, it is convenient instead to look at (4.12), which gives

$$\delta^2 L = 2\sqrt{-h}\left(h^{\alpha\beta}\delta^{kl}(D_{ki\alpha}\phi^i)(D_{lj\beta}\phi^j) - \frac{2}{h^{\epsilon\zeta}g_{\epsilon\zeta}}K^{\alpha\beta}_i K_{j\alpha\beta}\phi^i\phi^j\right) \tag{4.25}$$

where we introduced $D_{ij\alpha} = \delta_{ij}D_\alpha + \mu_{ij\alpha}$. Notice that the normal fundamental form plays the role of a gauge potential in this equation (compare with QED or QCD), while the second fundamental form acts as a scalar potential, or actually as a coordinate dependent mass-term.

The field equation corresponding to this Lagrangian is

$$\left(\delta^{kl}h^{\alpha\beta}D_{ik\alpha}D_{lj\beta} + \frac{2}{h^{\epsilon\zeta}g_{\epsilon\zeta}}K^{\alpha\beta}_i K_{j\alpha\beta}\right)\phi^j = 0 \tag{4.26}$$

Equations (4.26) are $D-2$ coupled linear partial differential equations with τ and σ dependent coefficients. Written out they read

$$\Box^{(h)}\phi_i + (D_\alpha\mu^{\alpha}_{ij})\phi^j + 2\mu^{\alpha}_{ij}\phi^j_{,\alpha} + \delta^{kl}\mu_{ik\alpha}\mu^{\alpha}_{lj}\phi^j + \frac{2}{h^{\epsilon\zeta}g_{\epsilon\zeta}}K^{\alpha\beta}_i K_{j\alpha\beta}\phi^j = 0 \tag{4.27}$$

where we used that $\Box^{(h)} = h^{\alpha\beta}D_\alpha D_\beta$, which is also given by equation (4.5).

4.2 Curved Spacetime

We now take a general spacetime metric $G_{\mu\nu}(X)$. This gives rise to a lot of extra terms with derivatives and second derivatives of the metric. These terms can be written using covariant derivatives and the Riemann tensor. For instance, the curved spacetime analog of (4.10) becomes (∇_ρ is the covariant derivative with respect to the metric $G_{\mu\nu}$)

$$\begin{aligned}&(\frac{1}{2}h^{\alpha\beta}h^{\gamma\delta} - h^{\alpha\gamma}h^{\beta\delta})h^{\epsilon\zeta}g_{\epsilon\zeta}\delta h_{\gamma\delta} \\ &= 2(\frac{1}{2}h^{\alpha\beta}h^{\gamma\delta} - h^{\alpha\gamma}h^{\beta\delta})G_{\mu\nu}(X^\nu_{,\gamma}(\nabla_\phi\delta X^\mu)X^\phi_{,\delta} + X^\nu_{,\delta}(\nabla_\phi\delta X^\mu)X^\phi_{,\gamma})\end{aligned} \tag{4.28}$$

while (4.12) turns into

$$\begin{aligned}\delta^2 L &= 2\frac{\sqrt{-h}}{h^{\epsilon\zeta}g_{\epsilon\zeta}}(\frac{1}{2}h^{\alpha\beta}h^{\gamma\delta} - h^{\alpha\gamma}h^{\beta\delta})G_{\mu\nu}G_{\rho\sigma} \\ &\quad\times(X^\nu_{,\gamma}(\nabla_\phi\delta X^\mu)X^\phi_{,\delta} + X^\nu_{,\delta}(\nabla_\phi\delta X^\mu)X^\phi_{,\gamma}) \\ &\quad\times(X^\sigma_{,\alpha}(\nabla_\psi\delta X^\rho)X^\psi_{,\beta} + X^\sigma_{,\beta}(\nabla_\psi\delta X^\rho)X^\psi_{,\alpha}) \\ &+ 2\sqrt{-h}h^{\alpha\beta}(G_{\mu\nu}(\nabla_\phi\delta X^\nu)X^\phi_{,\alpha}(\nabla_\psi\delta X^\mu)X^\psi_{,\beta} - R_{\mu\rho\nu\sigma}\delta X^\mu\delta X^\nu X^\rho_{,\alpha}X^\sigma_{,\beta})\end{aligned} \tag{4.29}$$

The rest of Subsection 4.1 is generalised similarly, so we shall skip the details. Equations (4.15), (4.16) obviously become

$$G_{\mu\nu}N^\mu_i N^\nu_j = \delta_{ij}, \quad G_{\mu\nu}N^\mu_i X^\nu_{,\alpha} = 0 \tag{4.30}$$

$$G^{\mu\nu} = \frac{2h^{\alpha\beta}}{h^{\gamma\delta}g_{\gamma\delta}}X^\mu_{,\alpha}X^\nu_{,\beta} + \delta^{ij}N^\mu_i N^\nu_j \tag{4.31}$$

while the definitions (4.17) now read

$$K^i_{\alpha\beta} = N^i_\mu X^\rho_{,\beta}\nabla_\rho X^\mu_{,\alpha}, \quad \mu^{ij}_\alpha = N^i_\mu X^\rho_{,\alpha}\nabla_\rho N^{j\mu} \tag{4.32}$$

Notice that we still have $K^i_{\alpha\beta} = K^i_{\beta\alpha}$ while $\mu^{ij}_\alpha = -\mu^{ji}_\alpha$.
The Weingarten equations become

$$X^\rho_{,\gamma}\nabla_\rho X^\mu_{,\beta} - \Gamma^\epsilon_{\beta\gamma}X^\mu_{,\epsilon} = N^\mu_i K^i_{\gamma\beta} + f^\delta_{\gamma\beta}X^\mu_{,\delta} \tag{4.33}$$

$$X^\rho_{,\alpha}\nabla_\rho N^{i\mu} = \mu^{ji}_\alpha N^\mu_j - \frac{2}{h^{\epsilon\zeta}g_{\epsilon\zeta}}K^{i\gamma}_\alpha X^\mu_{,\gamma} \tag{4.34}$$

where $f^\alpha_{\beta\gamma}$ is

$$X^\rho_{,\gamma}\nabla_\rho g_{\alpha\beta} - \Gamma^\epsilon_{\alpha\gamma}g_{\beta\epsilon} - \Gamma^\epsilon_{\beta\gamma}g_{\alpha\epsilon} = f^\delta_{\gamma\beta}g_{\alpha\delta} + f^\delta_{\gamma\alpha}g_{\beta\delta} \tag{4.35}$$

As for the perturbations, we still have the decomposition (4.21). The tangential perturbations correspond to symmetry transformations, as in Minkowski space. So we concentrate on the normal perturbations $\delta X^\mu = N^\mu_i\phi^i$. Then

$$X^\rho_{,\alpha}\nabla_\rho\delta X^\mu = X^\rho_{,\alpha}(\phi_i\nabla_\rho N^{\mu i} + N^\mu_i\nabla_\rho\phi^i) = \mu^{ji}_\alpha N^\mu_j\phi_i - \frac{2}{h^{\epsilon\zeta}g_{\epsilon\zeta}}K^{i\gamma}_\alpha X^\mu_{,\gamma}\phi_i + N^\mu_i\phi^i_{,\alpha} \tag{4.36}$$

i.e.

$$X^{\rho}_{,\alpha}\nabla_{\rho}\delta X^{\mu} = N^{\mu j}D_{ji\alpha}\phi^{i} - \frac{2}{h^{\epsilon\zeta}g_{\epsilon\zeta}}K^{i\gamma}_{\alpha}X^{\mu}_{,\gamma}\phi_{i} \tag{4.37}$$

Now (4.24) and (4.25) take the form

$$(\frac{1}{2}h^{\alpha\beta}h^{\gamma\delta} - h^{\alpha\gamma}h^{\beta\delta})\delta h_{\gamma\delta} = \frac{4}{h^{\epsilon\zeta}g_{\epsilon\zeta}}K^{i\alpha\beta}\phi_{i} \tag{4.38}$$

$$\begin{aligned}\delta^{2}L &= 2\sqrt{-h}(h^{\alpha\beta}\delta^{lk}(D_{ki\alpha}\phi^{i})(D_{lj\beta}\phi^{j}) - 2\frac{K^{i\alpha\beta}K^{j}_{\alpha\beta}}{h^{\epsilon\zeta}g_{\epsilon\zeta}}\phi_{i}\phi_{j} \\ &- h^{\alpha\beta}R_{\mu\rho\nu\sigma}N^{\mu}_{i}N^{\nu}_{j}X^{\rho}_{,\alpha}X^{\sigma}_{,\beta}\phi^{i}\phi^{j})\end{aligned} \tag{4.39}$$

The analog of (4.26) becomes

$$\left(\delta^{kl}h^{\alpha\beta}D_{ik\alpha}D_{lj\beta} + \frac{2}{h^{\epsilon\zeta}g_{\epsilon\zeta}}K^{\alpha\beta}_{i}K_{j\alpha\beta} + h^{\alpha\beta}R_{\mu\rho\nu\sigma}N^{\mu}_{i}N^{\nu}_{j}X^{\rho}_{,\alpha}X^{\sigma}_{,\beta}\right)\phi^{j} = 0 \tag{4.40}$$

which can also be written

$$\begin{aligned}&\Box^{(h)}\phi_{i} + (D_{\alpha}\mu^{\alpha}_{ij})\phi^{j} + 2\mu^{\alpha}_{ij}\phi^{j}_{,\alpha} + \delta^{kl}\mu_{ik\alpha}\mu^{\alpha}_{lj}\phi^{j} + \frac{2}{h^{\epsilon\zeta}g_{\epsilon\zeta}}K^{\alpha\beta}_{i}K_{j\alpha\beta}\phi^{j} \\ &+h^{\alpha\beta}R_{\mu\rho\nu\sigma}N^{\mu}_{i}N^{\nu}_{j}X^{\rho}_{,\alpha}X^{\sigma}_{,\beta}\phi^{j} = 0\end{aligned} \tag{4.41}$$

5 Stability Analysis

The circular pulsating string ansatz is well-known to be classically rigid in AdS. It is even more rigid than in Minkowski space since the gravitational field leads to further stabilisation (in de Sitter space it is the opposite). And this holds for both expanding and contracting strings. The perturbations around the circular pulsating string blow up when the string collapses in AdS_5 (as in Minkowski space), but the classical picture is not valid there. For the special case of perturbations around circular pulsating strings, see for instance [49].

As for circular spinning strings, the situation is quite different. They tend to be unstable for very large angular momentum. Thus our pulsating spinning string solitons of Section 3 will certainly be unstable for certain ranges of the parameters (for instance in the limit where they reduce to those condsidered by Tseytlin and Frolov (Subsection 2.3). Similarly, they will certainly be stable in other ranges (small amplitude and not very large angular momentum). Whether the coupling to pulsation more generally leads to stabilization or the contrary is an open problem, as we mentioned in our introduction section. It is unfortunately extremely difficult to settle this question. Comparing with the stability analysis by Tseytlin and Frolov [27], which corresponds to our solutions without pulsation: They had 3 equations with constant coefficients for the physical perturbations. When including pulsation we will have 8 equations with time dependent coefficients.

Anyway, the purpose of this section is precisely to settle that question using a combination of analytical and numerical methods. As a warm up, we start by determining the stability of the non-pulsating string considered in Subsection 2.3. This was already done in [27], using a different approach.

5.1 Stability of Two-Spin String Soliton

As the unperturbed string we take the two-spin solution in AdS_5, considered in Subsection 2.3. That is, $h_{\alpha\beta} = \eta_{\alpha\beta} = diag(-1,\ 1)$ as well as

$$X^\mu = (c_0\tau, r_0, \sigma, \omega\tau, \omega\tau) \tag{5.42}$$

where (c_0, r_0, ω) are constants. The tangent vectors are

$$\dot{X}^\mu = (c_0, 0, 0, \omega, \omega) \tag{5.43}$$

$$X'^\mu = (0, 0, 1, 0, 0) \tag{5.44}$$

where

$$\omega^2 = 1 + H^2 c_0^2 \tag{5.45}$$

$$c_0^2 = 2r_0^2 \tag{5.46}$$

The induced metric on the world-sheet is

$$\begin{aligned} g_{\tau\sigma} &= 0, \\ g_{\tau\tau} &= -r_0^2 \\ g_{\sigma\sigma} &= r_0^2 \end{aligned} \tag{5.47}$$

There are 3 normal vectors N^i $(i = 1, 2, 3)$ fulfilling eqs.(4.30)-(4.31). They can be chosen as

$$N^1_\mu = \frac{r_0\sqrt{1 + H^2 r_0^2}}{\sqrt{c_0^2 - r_0^2}}(\omega, 0, 0, -c_0 \sin^2\sigma, -c_0\cos^2\sigma) \tag{5.48}$$

$$N^2_\mu = r_0 \sin\sigma\cos\sigma(0, 0, 0, 1, -1) \tag{5.49}$$

$$N^3_\mu = (1 + H^2 r_0^2)^{-1/2}(0, 1, 0, 0, 0) \tag{5.50}$$

The normal vectors with upper-μ index are

$$N^{1\mu} = \frac{r_0\sqrt{1 + H^2 r_0^2}}{\sqrt{c_0^2 - r_0^2}}\left(-\frac{\omega}{1 + H^2 r_0^2}, 0, 0, \frac{-c_0}{r_0^2}, \frac{-c_0}{r_0^2}\right) \tag{5.51}$$

$$N^{2\mu} = \left(0, 0, 0, \frac{\cos\sigma}{r_0\sin\sigma}, \frac{-\sin\sigma}{r_0\cos\sigma}\right) \tag{5.52}$$

$$N^{3\mu} = \sqrt{1 + H^2 r_0^2}\,(0, 1, 0, 0, 0) \tag{5.53}$$

The second fundamental form and normal fundamental form give (we only show the non-zero components)

$$K^3_{\tau\tau} = K^3_{\sigma\sigma} = -r_0\sqrt{1+H^2r_0^2} \tag{5.54}$$

$$K^2_{\tau\sigma} = \omega r_0 \tag{5.55}$$

$$\mu^{21}_{\sigma} = -\mu^{12}_{\sigma} = \frac{-c_0\sqrt{1+H^2r_0^2}}{\sqrt{c_0^2-r_0^2}} \tag{5.56}$$

$$\mu^{31}_{\tau} = -\mu^{13}_{\tau} = \frac{c_0\omega}{\sqrt{c_0^2-r_0^2}} \tag{5.57}$$

We also have to compute

$$h^{\alpha\beta}R_{\mu\rho\nu\sigma}N^{\mu}_{i}N^{\nu}_{j}X^{\rho}_{,\alpha}X^{\sigma}_{,\beta} \tag{5.58}$$

which in our gauge becomes

$$R_{\mu\rho\nu\sigma}N^{\mu}_{i}N^{\nu}_{j}(X'^{\rho}X'^{\sigma}-\dot{X}^{\rho}\dot{X}^{\sigma}) \tag{5.59}$$

which leads to

$$R_{\mu\rho\nu\sigma}N^{\mu}_{1}N^{\nu}_{1}(X'^{\rho}X'^{\sigma}-\dot{X}^{\rho}\dot{X}^{\sigma}) = -H^2c_0^2 \tag{5.60}$$

$$R_{\mu\rho\nu\sigma}N^{\mu}_{2}N^{\nu}_{2}(X'^{\rho}X'^{\sigma}-\dot{X}^{\rho}\dot{X}^{\sigma}) = -H^2c_0^2 \tag{5.61}$$

$$R_{\mu\rho\nu\sigma}N^{\mu}_{3}N^{\nu}_{3}(X'^{\rho}X'^{\sigma}-\dot{X}^{\rho}\dot{X}^{\sigma}) = -H^2c_0^2 \tag{5.62}$$

We now have all the ingredients to write down the equations (4.40) for the perturbations

$$-\ddot{\phi}^1+\phi''^1+\sqrt{8(1+H^2c_0^2)}\dot{\phi}^3+\sqrt{4(2+H^2c_0^2)}\phi'^2=0 \tag{5.63}$$

$$-\ddot{\phi}^2+\phi''^2-\sqrt{4(2+H^2c_0^2)}\phi'^1-4(1+H^2c_0^2)\phi^2=0 \tag{5.64}$$

$$-\ddot{\phi}^3+\phi''^3-\sqrt{8(1+H^2c_0^2)}\dot{\phi}^1+2(2+H^2c_0^2)\phi^3=0 \tag{5.65}$$

When we have solved these equations, the coordinate perturbations are given by $\delta X^{\mu} = N^{\mu}_{i}\phi^{i}$

$$\delta t = \frac{-r_0\omega}{\sqrt{c_0^2-r_0^2}\sqrt{1+H^2r_0^2}}\phi^1 \tag{5.66}$$

$$\delta r = \sqrt{1+H^2r_0^2}\,\phi^3 \tag{5.67}$$

$$\delta\beta = 0 \tag{5.68}$$

$$\delta\phi = \frac{-c_0\sqrt{1+H^2r_0^2}}{r_0\sqrt{c_0^2-r_0^2}}\phi^1+\frac{\cos\sigma}{r_0\sin\sigma}\phi^2 \tag{5.69}$$

$$\delta\tilde{\phi} = \frac{-c_0\sqrt{1+H^2r_0^2}}{r_0\sqrt{c_0^2-r_0^2}}\phi^1-\frac{\sin\sigma}{r_0\cos\sigma}\phi^2 \tag{5.70}$$

These equations can be inverted to

$$\begin{aligned}
\phi^1 &= -\frac{\sqrt{c_0^2 - r_0^2}\sqrt{1 + H^2 r_0^2}}{r_0 \omega}\delta t & (5.71)\\
\phi^2 &= r_0 \cos\sigma \sin\sigma(\delta\phi - \delta\tilde{\phi}) & (5.72)\\
\phi^3 &= \frac{1}{\sqrt{1 + H^2 r_0^2}}\delta r & (5.73)
\end{aligned}$$

Provided that the consistency relation is fulfilled

$$\frac{c_0(1 + H^2 r_0^2)}{r_0^2 \omega}\delta t - \cos^2\sigma\delta\tilde{\phi} - \sin^2\sigma\delta\phi = 0 \tag{5.74}$$

which in some sense defines our gauge choice for the perturbations. For comparison we notice that Frolov and Tseytlin [27] used the gauge $\delta t = 0$.

To solve eqs.(5.63)-(5.65), we make Fourier expansions

$$\phi^i(\tau, \sigma) = \sum_n e^{in\sigma}\phi_n^i(\tau) \tag{5.75}$$

where ${\phi_n^i}^* = \phi_{-n}^i$. We also turn the second order equations into first order ones by defining $p_n^i = \dot{\phi}_n^i$, and we introduce the dimensionless parameter $\kappa = Hc_0$. Then we get

$$\begin{pmatrix} \dot{\phi}_n^1 \\ \dot{p}_n^1 \\ \dot{\phi}_n^2 \\ \dot{p}_n^2 \\ \dot{\phi}_n^3 \\ \dot{p}_n^3 \end{pmatrix} + \mathcal{A}\begin{pmatrix} \phi_n^1 \\ p_n^1 \\ \phi_n^2 \\ p_n^2 \\ \phi_n^3 \\ p_n^3 \end{pmatrix} = \begin{pmatrix} 0 \\ 0 \\ 0 \\ 0 \\ 0 \\ 0 \end{pmatrix} \tag{5.76}$$

where the matrix $\mathcal{A}$ is given by

$$\begin{pmatrix}
0 & -1 & 0 & 0 & 0 & 0 \\
n^2 & 0 & -2in\sqrt{2+\kappa^2} & 0 & 0 & -2\sqrt{2(1+\kappa^2)} \\
0 & 0 & 0 & -1 & 0 & 0 \\
2in\sqrt{2+\kappa^2} & 0 & n^2+4(1+\kappa^2) & 0 & 0 & 0 \\
0 & 0 & 0 & 0 & 0 & -1 \\
0 & 2\sqrt{2(1+\kappa^2)} & 0 & 0 & n^2-2(2+\kappa^2) & 0
\end{pmatrix} \tag{5.77}$$

To solve the system (5.76), we recall a simple theorem from linear algebra. Consider a first order matrix system of differential equations

$$\dot{\vec{v}}(\tau) + \mathcal{A}\vec{v}(\tau) = 0 \tag{5.78}$$

where $\mathcal{A}$ is a constant matrix. The solution is given by

$$\vec{v}(\tau) = e^{-\mathcal{A}\tau}\vec{v}(\tau_0) \tag{5.79}$$

The matrix $e^{-\mathcal{A}\tau}$ can be computed using the power expansion for the exponential function. However, for our purposes it is enough to use that all components will be of the form

$$\sum_{\lambda} e^{\alpha\tau}(P_\lambda(\tau)\cos(\beta\tau) + Q_\lambda(\tau)\sin(\beta\tau)) \tag{5.80}$$

where $\lambda = \alpha + i\beta$ are the eigenvalues of $\mathcal{A}$ and $(P_\lambda,\ Q_\lambda)$ are polynomials in τ, which appear if some eigenvalues have multiplicity.

Defining $l = \lambda^2$, the characteristic polynomial of (5.77) is

$$\begin{aligned} p(l) &= l^3 + l^2(3n^2 + 10\kappa^2 + 8) + l(3n^4 + 24\kappa^4 + 40\kappa^2 + 8\kappa^2 n^2 + 16) \\ &+ n^6 - 2n^4\kappa^2 - 8n^4 + 8n^2\kappa^2 + 16n^2 \end{aligned} \tag{5.81}$$

In order for all three roots to be negative and thus λ imaginary, we need that both extrema fall at negative l (which is easy to show) and that $p(0) \geq 0$. This leads to

$$(n^2 - 4)(n^4 - 4n^2 - 2n^2\kappa^2) \geq 0 \tag{5.82}$$

This means that for $n = 0, 1, 2$ κ is arbitrary. For $n = 3, 4, 5$ we get $\kappa \leq \sqrt{\frac{5}{2}}, \sqrt{6}, \sqrt{\frac{21}{2}}$, respectively. For larger n it is easy to see that we get weaker conditions, so we end up with $\kappa^2 \leq 5/2$.

When this condition is fulfilled, all the perturbations are oscillating and thus we have stability. The polynomials P_λ and Q_λ in this picture represent the various zero-modes which can be absorbed in the unperturbed solution.

The result $\kappa^2 \leq 5/2$ is of course in agreement with [27].

5.2 Stability of Pulsating Two-Spin Soliton

As the unperturbed string we take our pulsating two-spin solution in $AdS_5 \times S^5$. That is, $h_{\alpha\beta} = \eta_{\alpha\beta} = diag(-1,\ 1)$ as well as

$$X^\mu = (c_0\tau, r_0, \sigma, \omega\tau, \omega\tau, \theta(\tau), \sigma, \psi_{10}, \psi_{20}, \psi_{30}) \tag{5.83}$$

where $(c_0, r_0, \omega, \psi_{10}, \psi_{20}, \psi_{30})$ are constants. The tangent vectors are

$$\dot{X}^\mu = (c_0, 0, 0, \omega, \omega, \dot{\theta}, 0, 0, 0, 0) \tag{5.84}$$

$$X'^\mu = (0, 0, 1, 0, 0, 0, 1, 0, 0, 0) \tag{5.85}$$

where

$$\omega^2 = 1 + H^2 c_0^2 \tag{5.86}$$

$$\dot{\theta}^2 + \sin^2\theta = H^2(c_0^2 - 2r_0^2) \tag{5.87}$$

The induced metric on the world-sheet is

$$\begin{aligned} g_{\tau\sigma} &= 0, \\ g_{\tau\tau} &= -H^{-2}(H^2 r_0^2 + \sin^2\theta) \\ g_{\sigma\sigma} &= H^{-2}(H^2 r_0^2 + \sin^2\theta) \end{aligned} \tag{5.88}$$

There are 8 normal vectors N_μ^i $(i = 1, 2, ..., 8)$ fulfilling eqs.(4.30)-(4.31). They can be chosen as

$$N_\mu^1 = \frac{r_0\sqrt{1 + H^2 r_0^2}}{\sqrt{c_0^2 - r_0^2}}(\omega, 0, 0, -c_0 \sin^2\sigma, -c_0\cos^2\sigma, 0, 0, 0, 0, 0) \tag{5.89}$$

$$N_\mu^2 = r_0 \sin\sigma\cos\sigma(0, 0, 0, 1, -1, 0, 0, 0, 0, 0) \tag{5.90}$$

$$N_\mu^3 = (1 + H^2 r_0^2)^{-1/2}(0, 1, 0, 0, 0, 0, 0, 0, 0, 0) \tag{5.91}$$

$$N_\mu^4 = \frac{r_0 \sin\theta}{\sqrt{H^2 r_0^2 + \sin^2\theta}}(0, 0, 1, 0, 0, 0, -1, 0, 0, 0) \tag{5.92}$$

$$N_\mu^5 = \frac{1}{\sqrt{c_0^2 - r_0^2}\sqrt{H^2 r_0^2 + \sin^2\theta}} \tag{5.93}$$

$$\left(-c_0(1 + H^2 r_0^2)\dot{\theta}, 0, 0, \omega r_0^2 \sin^2\sigma\dot{\theta}, \omega r_0^2\cos^2\sigma\dot{\theta}, c_0^2 - r_0^2, 0, 0, 0, 0\right)$$

$$N_\mu^6 = H^{-1}\cos\theta(0, 0, 0, 0, 0, 0, 0, 1, 0, 0) \tag{5.94}$$

$$N_\mu^7 = H^{-1}\cos\theta\sin\psi_{10}(0, 0, 0, 0, 0, 0, 0, 0, 1, 0) \tag{5.95}$$

$$N_\mu^8 = H^{-1}\cos\theta\cos\psi_{10}(0, 0, 0, 0, 0, 0, 0, 0, 0, 1) \tag{5.96}$$

Notice that the first 3 only have components in AdS_5 while the last 3 only have components in S^5. The normal vectors with upper-μ index are

$$N^{1\mu} = \frac{r_0\sqrt{1 + H^2 r_0^2}}{\sqrt{c_0^2 - r_0^2}}(-\frac{\omega}{1 + H^2 r_0^2}, 0, 0, \frac{-c_0}{r_0^2}, \frac{-c_0}{r_0^2}, 0, 0, 0, 0, 0) \tag{5.97}$$

$$N^{2\mu} = (0, 0, 0, \frac{\cos\sigma}{r_0\sin\sigma}, \frac{-\sin\sigma}{r_0\cos\sigma}, 0, 0, 0, 0, 0) \tag{5.98}$$

$$N^{3\mu} = \sqrt{1+H^2r_0^2}(0,1,0,0,0,0,0,0,0,0) \tag{5.99}$$

$$N^{4\mu} = \frac{r_0\sin\theta}{\sqrt{H^2r_0^2+\sin^2\theta}}(0,0,\frac{1}{r_0^2},0,0,0,\frac{-H^2}{\sin^2\theta},0,0,0) \tag{5.100}$$

$$N^{5\mu} = \frac{1}{\sqrt{c_0^2-r_0^2}\sqrt{H^2r_0^2+\sin^2\theta}} \left(c_0\dot{\theta},0,0,\omega\theta,\omega\dot{\theta},H^2(c_0^2-r_0^2),0,0,0,0\right) \tag{5.101}$$

$$N^{6\mu} = \frac{H}{\cos\theta}(0,0,0,0,0,0,0,1,0,0) \tag{5.102}$$

$$N^{7\mu} = \frac{H}{\cos\theta\sin\psi_{10}}(0,0,0,0,0,0,0,0,1,0) \tag{5.103}$$

$$N^{8\mu} = \frac{H}{\cos\theta\cos\psi_{10}}(0,0,0,0,0,0,0,0,0,1) \tag{5.104}$$

The second fundamental form and normal fundamental form give

$$K^3_{\tau\tau} = K^3_{\sigma\sigma} = -r_0\sqrt{1+H^2r_0^2} \tag{5.105}$$

$$K^5_{\tau\tau} = K^5_{\sigma\sigma} = \frac{-\sqrt{c_0^2-r_0^2}\,\sin\theta\cos\theta}{\sqrt{H^2r_0^2+\sin^2\theta}} \tag{5.106}$$

$$K^2_{\tau\sigma} = \omega r_0 \tag{5.107}$$

$$K^4_{\tau\sigma} = \frac{-r_0\cos\theta\dot{\theta}}{\sqrt{H^2r_0^2+\sin^2\theta}} \tag{5.108}$$

$$\mu^{21}_{\sigma} = -\mu^{12}_{\sigma} = \frac{-c_0\sqrt{1+H^2r_0^2}}{\sqrt{c_0^2-r_0^2}} \tag{5.109}$$

$$\mu^{25}_{\sigma} = -\mu^{52}_{\sigma} = \frac{r_0\omega\dot{\theta}}{\sqrt{c_0^2-r_0^2}\sqrt{H^2r_0^2+\sin^2\theta}} \tag{5.110}$$

$$\mu^{34}_{\sigma} = -\mu^{43}_{\sigma} = \frac{-\sqrt{1+H^2r_0^2}\,\sin\theta}{\sqrt{H^2r_0^2+\sin^2\theta}} \tag{5.111}$$

$$\mu^{45}_{\sigma} = -\mu^{54}_{\sigma} = \frac{-H^2r_0\cos\theta\sqrt{c_0^2-r_0^2}}{H^2r_0^2+\sin^2\theta} \tag{5.112}$$

$$\mu^{31}_{\tau} = -\mu^{13}_{\tau} = \frac{c_0\omega}{\sqrt{c_0^2-r_0^2}} \tag{5.113}$$

$$\mu_\tau^{42} = -\mu_\tau^{24} = \frac{-\omega \sin\theta}{\sqrt{H^2 r_0^2 + \sin^2\theta}} \tag{5.114}$$

$$\mu_\tau^{53} = -\mu_\tau^{35} = \frac{r_0 \dot{\theta}\sqrt{1 + H^2 r_0^2}}{\sqrt{c_0^2 - r_0^2}\sqrt{H^2 r_0^2 + \sin^2\theta}} \tag{5.115}$$

Now the Riemann component terms

$$R_{\mu\rho\nu\sigma} N_i^\mu N_j^\nu (X'^\rho X'^\sigma - \dot{X}^\rho \dot{X}^\sigma) \tag{5.116}$$

lead to

$$\begin{aligned}
R_{\mu\rho\nu\sigma} N_1^\mu N_1^\nu (X'^\rho X'^\sigma - \dot{X}^\rho \dot{X}^\sigma) &= -H^2 c_0^2 && (5.117)\\
R_{\mu\rho\nu\sigma} N_2^\mu N_2^\nu (X'^\rho X'^\sigma - \dot{X}^\rho \dot{X}^\sigma) &= -H^2 c_0^2 && (5.118)\\
R_{\mu\rho\nu\sigma} N_3^\mu N_3^\nu (X'^\rho X'^\sigma - \dot{X}^\rho \dot{X}^\sigma) &= -H^2 c_0^2 && (5.119)\\
R_{\mu\rho\nu\sigma} N_4^\mu N_4^\nu (X'^\rho X'^\sigma - \dot{X}^\rho \dot{X}^\sigma) &= H^2 (2r_0^2 - c_0^2) && (5.120)\\
R_{\mu\rho\nu\sigma} N_5^\mu N_5^\nu (X'^\rho X'^\sigma - \dot{X}^\rho \dot{X}^\sigma) &= -H^2 (2r_0^2 - c_0^2) && (5.121)\\
R_{\mu\rho\nu\sigma} N_6^\mu N_6^\nu (X'^\rho X'^\sigma - \dot{X}^\rho \dot{X}^\sigma) &= 2\sin^2\theta - H^2 (c_0^2 - 2r_0^2) && (5.122)\\
R_{\mu\rho\nu\sigma} N_7^\mu N_7^\nu (X'^\rho X'^\sigma - \dot{X}^\rho \dot{X}^\sigma) &= 2\sin^2\theta - H^2 (c_0^2 - 2r_0^2) && (5.123)\\
R_{\mu\rho\nu\sigma} N_8^\mu N_8^\nu (X'^\rho X'^\sigma - \dot{X}^\rho \dot{X}^\sigma) &= 2\sin^2\theta - H^2 (c_0^2 - 2r_0^2) && (5.124)
\end{aligned}$$

We now have all the ingredients to write down the equations (4.40) for the perturbations

$$-\ddot{\phi}^1 + \phi''^1 + \frac{2c_0\omega}{\sqrt{c_0^2 - r^2}}\dot{\phi}^3 + \frac{2c_0\sqrt{1 + H^2 r^2}}{\sqrt{c_0^2 - r^2}}\phi'^2 = 0 \tag{5.125}$$

$$\begin{aligned}
&-\ddot{\phi}^2 + \phi''^2 - \frac{2\omega\sin\theta}{\sqrt{H^2 r^2 + \sin^2\theta}}\dot{\phi}^4 - \frac{2c_0\sqrt{1 + H^2 r^2}}{\sqrt{c_0^2 - r^2}}\phi'^1\\
&+ \frac{2r\omega\dot{\theta}}{\sqrt{c_0^2 - r^2}\sqrt{H^2 r^2 + \sin^2\theta}}\phi'^5 - \frac{4H^2 r^2 (H^2 c_0^2 + 1)}{H^2 r^2 + \sin^2\theta}\phi^2\\
&+ \frac{2H^2 r^2 \omega\cos\theta\dot{\theta}}{(H^2 r^2 + \sin^2\theta)^{3/2}}\phi^4 = 0
\end{aligned} \tag{5.126}$$

$$
\begin{aligned}
&-\ddot{\phi}^3 + \phi''^3 - \frac{2c_0\omega}{\sqrt{c_0^2 - r^2}}\,\dot{\phi}^1 + \frac{2r\dot{\theta}\sqrt{1+H^2r^2}}{\sqrt{c_0^2-r^2}\sqrt{H^2r^2+\sin^2\theta}}\,\dot{\phi}^5 \\
&- \frac{2\sqrt{1+H^2r^2}\sin\theta}{\sqrt{H^2r^2+\sin^2\theta}}\,\phi'^4 + \frac{4H^2r^2(1+H^2r^2)}{H^2r^2+\sin^2\theta}\,\phi^3 \\
&+ \frac{2H^2r\cos\theta\sin\theta\sqrt{1+H^2r^2}\sqrt{c_0^2-r^2}}{(H^2r^2+\sin^2\theta)^{3/2}}\,\phi^5 = 0
\end{aligned} \tag{5.127}
$$

$$
\begin{aligned}
&-\ddot{\phi}^4 + \phi''^4 + \frac{2\omega\sin\theta}{\sqrt{H^2r^2+\sin^2\theta}}\,\dot{\phi}^2 + \frac{2\sqrt{1+H^2r^2}\sin\theta}{\sqrt{H^2r^2+\sin^2\theta}}\,\phi'^3 \\
&- \frac{2H^2r\cos\theta\sqrt{c_0^2-r^2}}{H^2r^2+\sin^2\theta}\,\phi'^5 + \frac{4H^2r^2\omega\dot{\theta}\cos\theta}{(H^2r^2+\sin^2\theta)^{3/2}}\,\phi^2 \\
&+ \frac{H^2r^2[-3H^2c_0^2+5H^2r^2+2H^4r^4-H^4c_0^2r^2]}{(H^2r^2+\sin^2\theta)^2}\,\phi^4 \\
&+ \frac{H^2r^2[(2-2H^2r^2+2H^2c_0^2)\sin^2\theta-\sin^4\theta]}{(H^2r^2+\sin^2\theta)^2}\,\phi^4 = 0
\end{aligned} \tag{5.128}
$$

$$
\begin{aligned}
&-\ddot{\phi}^5 + \phi''^5 - \frac{2r\dot{\theta}\sqrt{1+H^2r^2}}{\sqrt{c_0^2-r^2}\sqrt{H^2r^2+\sin^2\theta}}\,\dot{\phi}^3 \\
&- \frac{2r\omega\dot{\theta}}{\sqrt{c_0^2-r^2}\sqrt{H^2r^2+\sin^2\theta}}\,\phi'^2 + \frac{2H^2r\cos\theta\sqrt{c_0^2-r^2}}{H^2r^2+\sin^2\theta}\,\phi'^4 \\
&+ \frac{4H^2r\sin\theta\cos\theta\sqrt{c_0^2-r^2}\sqrt{1+H^2r^2}}{(H^2r^2+\sin^2\theta)^{3/2}}\phi^3 \\
&+ \frac{H^2(c_0^2-r^2)(-H^2r^2+2(H^2r^2+1)\sin^2\theta-\sin^4\theta)}{(H^2r^2+\sin^2\theta)^2}\phi^5 = 0
\end{aligned} \tag{5.129}
$$

$$
-\ddot{\phi}^6 + \phi''^6 + (2\sin^2\theta - H^2(c_0^2-2r^2))\phi^6 = 0 \tag{5.130}
$$

$$
-\ddot{\phi}^7 + \phi''^7 + (2\sin^2\theta - H^2(c_0^2-2r^2))\phi^7 = 0 \tag{5.131}
$$

$$
-\ddot{\phi}^8 + \phi''^8 + (2\sin^2\theta - H^2(c_0^2-2r^2))\phi^8 = 0 \tag{5.132}
$$

When we have solved these equations (analytically or numerically), the coordinate perturbations are given by $\delta X^\mu - N_i^\mu\phi^i$, i.e.

$$
\begin{aligned}
\delta t &= \frac{-r_0\omega}{\sqrt{c_0^2-r_0^2}\sqrt{1+H^2r_0^2}}\phi^1 \\
&+ \frac{c_0\dot{\theta}}{\sqrt{c_0^2-r_0^2}\sqrt{H^2r_0^2+\sin^2\theta}}\phi^5
\end{aligned} \tag{5.133}
$$

$$
\delta r = \sqrt{1+H^2r_0^2}\,\phi^3 \tag{5.134}
$$

$$\delta\beta = \frac{\sin\theta}{r_0\sqrt{H^2 r_0^2 + \sin^2\theta}}\phi^4 \tag{5.135}$$

$$\begin{aligned}\delta\phi &= \frac{-c_0\sqrt{1+H^2r_0^2}}{r_0\sqrt{c_0^2-r_0^2}}\phi^1 + \frac{\cos\sigma}{r_0\sin\sigma}\phi^2 \\ &+ \frac{\omega\dot{\theta}}{\sqrt{c_0^2-r_0^2}\sqrt{H^2r_0^2+\sin^2\theta}}\phi^5\end{aligned} \tag{5.136}$$

$$\begin{aligned}\delta\tilde{\phi} &= \frac{-c_0\sqrt{1+H^2r_0^2}}{r_0\sqrt{c_0^2-r_0^2}}\phi^1 - \frac{\sin\sigma}{r_0\cos\sigma}\phi^2 \\ &+ \frac{\omega\dot{\theta}}{\sqrt{c_0^2-r_0^2}\sqrt{H^2r_0^2+\sin^2\theta}}\phi^5\end{aligned} \tag{5.137}$$

$$\delta\theta = \frac{H^2\sqrt{c_0^2-r_0^2}}{\sqrt{H^2r_0^2+\sin^2\theta}}\phi^5 \tag{5.138}$$

$$\delta\psi = \frac{-H^2 r_0}{\sin\theta\sqrt{H^2r_0^2+\sin^2\theta}}\phi^4 \tag{5.139}$$

$$\delta\psi_1 = \frac{H}{\cos\theta}\phi^6 \tag{5.140}$$

$$\delta\psi_2 = \frac{H}{\cos\theta\sin\psi_{10}}\phi^7 \tag{5.141}$$

$$\delta\psi_3 = \frac{H}{\cos\theta\sin\psi_{10}}\phi^8 \tag{5.142}$$

These equations can be inverted to

$$\phi^1 = \frac{c_0\dot{\theta}\sqrt{1+H^2r_0^2}}{H^2r_0\omega\sqrt{c_0^2-r_0^2}}\delta\theta - \frac{\sqrt{c_0^2-r_0^2}\sqrt{1+H^2r_0^2}}{r_0\omega}\delta t \tag{5.143}$$

$$\phi^2 = r_0\cos\sigma\sin\sigma(\delta\phi - \delta\tilde{\phi}) \tag{5.144}$$

$$\phi^3 = \frac{1}{\sqrt{1+H^2r_0^2}}\delta r \tag{5.145}$$

$$\phi^4 = \frac{-\sin\theta\sqrt{H^2r_0^2+\sin^2\theta}}{H^2 r_0}\delta\psi \tag{5.146}$$

$$\phi^5 = \frac{\sqrt{H^2r_0^2+\sin^2\theta}}{H^2\sqrt{c_0^2-r_0^2}}\delta\theta \tag{5.147}$$

$$
\begin{aligned}
\phi^6 &= \frac{\cos\theta}{H}\delta\psi_1 && (5.148)\\
\phi^7 &= \frac{\cos\theta\sin\psi_{10}}{H}\delta\psi_2 && (5.149)\\
\phi^8 &= \frac{\cos\theta\cos\psi_{10}}{H}\delta\psi_3 && (5.150)
\end{aligned}
$$

Provided that the 2 consistency relations are fulfilled

$$\delta\beta + \frac{\sin^2\theta}{H^2 r_0^2}\delta\psi = 0 \qquad (5.151)$$

$$\frac{c_0(1+H^2r_0^2)}{r_0^2\omega}\delta t - \frac{\dot{\theta}}{H^2 r_0^2\omega}\delta\theta - \cos^2\sigma\delta\tilde{\phi} - \sin^2\sigma\delta\phi = 0 \qquad (5.152)$$

which in some sense define our gauge choice for the perturbations.

To solve the equations (5.125)-(5.132), we proceed as in Subsection 5.1 and make Fourier expansions (5.75). We also insert the explicit solution (3.6) and define $\kappa = Hc_0$. Then we get for $A \leq 1$

$$
\begin{aligned}
&\ddot{\phi}_n^1 + n^2\phi_n^1 - \sqrt{\frac{8\kappa^2(1+\kappa^2)}{\kappa^2+A^2}}\dot{\phi}_n^3\\
&-in\sqrt{\frac{4\kappa^2(2+\kappa^2-A^2)}{\kappa^2+A^2}}\phi_n^2 = 0 \qquad (5.153)
\end{aligned}
$$

$$
\begin{aligned}
&\ddot{\phi}_n^2 + n^2\phi_n^2 + \sqrt{\frac{8(1+\kappa^2)}{\kappa^2 - A^2 + 2A^2\mathrm{sn}^2(\tau|A^2)}}A\mathrm{sn}(\tau|A^2)\dot{\phi}_n^4\\
&+in\sqrt{\frac{4\kappa^2(2+\kappa^2-A^2)}{\kappa^2+A^2}}\phi_n^1 + \frac{4(\kappa^2-A^2)(1+\kappa^2)}{\kappa^2 - A^2 + 2A^2\mathrm{sn}^2(\tau|A^2)}\phi_n^2\\
&-in\sqrt{\frac{8(\kappa^2-A^2)(1+\kappa^2)}{(\kappa^2+A^2)(\kappa^2 - A^2 + 2A^2\mathrm{sn}^2(\tau|A^2))}}A\mathrm{cn}(\tau|A^2)\phi_n^5\\
&-\sqrt{\frac{8(\kappa^2-A^2)^2(1+\kappa^2)}{(\kappa^2 - A^2 + 2A^2\mathrm{sn}^2(\tau|A^2))^3}}A\mathrm{dn}(\tau|A^2)\mathrm{cn}(\tau|A^2)\phi_n^4 = 0 \qquad (5.154)
\end{aligned}
$$

$$\begin{aligned}
&\ddot{\phi}^3_n + n^2\phi^3_n + \sqrt{\frac{8\kappa^2(1+\kappa^2)}{\kappa^2+A^2}}\dot{\phi}^1_n - \frac{2(\kappa^2-A^2)(2+\kappa^2-A^2)}{\kappa^2-A^2+2A^2\mathrm{sn}^2(\tau|A^2)}\phi^3_n \\
&-\sqrt{\frac{4(\kappa^2-A^2)(2+\kappa^2-A^2)}{(\kappa^2+A^2)(\kappa^2-A^2+2A^2\mathrm{sn}^2(\tau|A^2))}}A\mathrm{cn}(\tau|A^2)\dot{\phi}^5_n \\
&+in\sqrt{\frac{4(2+\kappa^2-A^2)}{\kappa^2-A^2+2A^2\mathrm{sn}^2(\tau|A^2)}}A\mathrm{sn}(\tau|A^2)\phi^4_n \\
&-\sqrt{\frac{4(\kappa^2-A^2)(\kappa^2+A^2)(2+\kappa^2-A^2)}{(\kappa^2-A^2+2A^2\mathrm{sn}^2(\tau|A^2))^3}}A\mathrm{dn}(\tau|A^2)\mathrm{sn}(\tau|A^2)\phi^5_n = 0
\end{aligned} \tag{5.155}$$

$$\begin{aligned}
&\ddot{\phi}^4_n + n^2\phi^4_n - \sqrt{\frac{8(1+\kappa^2)}{\kappa^2-A^2+2A^2\mathrm{sn}^2(\tau|A^2)}}A\mathrm{sn}(\tau|A^2)\dot{\phi}^2_n \\
&-in\sqrt{\frac{4(2+\kappa^2-A^2)}{\kappa^2-A^2+2A^2\mathrm{sn}^2(\tau|A^2)}}A\mathrm{sn}(\tau|A^2)\phi^3_n \\
&+in\sqrt{\frac{4(\kappa^2-A^2)(\kappa^2+A^2)}{(\kappa^2-A^2+2A^2\mathrm{sn}^2(\tau|A^2))^2}}\mathrm{dn}(\tau|A^2)\phi^5_n \\
&-\sqrt{\frac{32(\kappa^2-A^2)^2(1+\kappa^2)}{(\kappa^2-A^2+2A^2\mathrm{sn}^2(\tau|A^2))^3}}A\mathrm{cn}(\tau|A^2)\mathrm{dn}(\tau|A^2)\phi^2_n \\
&-\frac{2(\kappa^2-A^2)[-\frac{\kappa^2}{2}-\frac{\kappa^2A^2}{2}-\frac{5A^2}{2}+\frac{A^4}{2}]}{(\kappa^2-A^2+2A^2\mathrm{sn}^2(\tau|A^2))^2}\phi^4_n \\
&-\frac{2(\kappa^2-A^2)[(2+A^2+\kappa^2)A^2\mathrm{sn}^2(\tau|A^2)-A^4\mathrm{sn}^4(\tau|A^2)]}{(\kappa^2-A^2+2A^2\mathrm{sn}^2(\tau|A^2))^2}\phi^4_n = 0
\end{aligned} \tag{5.156}$$

$$\begin{aligned}
&\ddot{\phi}^5_n + n^2\phi^5_n + \sqrt{\frac{4(\kappa^2-A^2)(2+\kappa^2-A^2)}{(\kappa^2+A^2)(\kappa^2-A^2+2A^2\mathrm{sn}^2(\tau|A^2))}}A\mathrm{cn}(\tau|A^2)\dot{\phi}^3_n \\
&+in\sqrt{\frac{8(\kappa^2-A^2)(1+\kappa^2)}{(\kappa^2+A^2)(\kappa^2-A^2+2A^2\mathrm{sn}^2(\tau|A^2))}}A\mathrm{cn}(\tau|A^2)\phi^2_n \\
&-\sqrt{\frac{4(\kappa^2-A^2)(\kappa^2+A^2)}{(\kappa^2-A^2+2A^2\mathrm{sn}^2(\tau|A^2))^2}}\mathrm{dn}(\tau|A^2)\phi^4_n \\
&-\sqrt{\frac{16(\kappa^2-A^2)(\kappa^2+A^2)(2+\kappa^2-A^2)}{(\kappa^2-A^2+2A^2\mathrm{sn}^2(\tau|A^2))^3}}A\mathrm{sn}(\tau|A^2)\mathrm{dn}(\tau|A^2)\phi^3_n \\
&-\frac{(\kappa^2+A^2)[-(\kappa^2-A^2)+2(\kappa^2-A^2+2)A^2\mathrm{sn}^2(\tau|A^2)]}{(\kappa^2-A^2+2A^2\mathrm{sn}^2(\tau|A^2))^2}\phi^5_n \\
&+2A^4\frac{(\kappa^2+A^2)\mathrm{sn}^4(\tau|A^2)}{(\kappa^2-A^2+2A^2\mathrm{sn}^2(\tau|A^2))^2}\phi^5_n = 0
\end{aligned} \tag{5.157}$$

$$\ddot{\phi}^6_n + n^2\phi^6_n - (2\text{sn}^2(\tau|A^2) - 1)A^2\phi^6_n = 0 \quad (5.158)$$
$$\ddot{\phi}^7_n + n^2\phi^7_n - (2\text{sn}^2(\tau|A^2) - 1)A^2\phi^7_n = 0 \quad (5.159)$$
$$\ddot{\phi}^8_n + n^2\phi^8_n - (2\text{sn}^2(\tau|A^2) - 1)A^2\phi^8_n = 0 \quad (5.160)$$

For $A \geq 1$, the equations look similar, thus we do not write them down here.

5.3 Analytical Results

First we can look at the equations in some special limits. For $A = 0$, corresponding to no pulsation, we get

$$\ddot{\phi}^1_n + n^2\phi^1_n - \sqrt{8(1+\kappa^2)}\dot{\phi}^3_n - \sqrt{4(2+\kappa^2)}in\phi^2_n = 0 \quad (5.161)$$
$$\ddot{\phi}^2_n + n^2\phi^2_n + \sqrt{4(2+\kappa^2)}in\phi^1_n + 4(1+\kappa^2)\phi^2_n = 0 \quad (5.162)$$
$$\ddot{\phi}^3_n + n^2\phi^3_n + \sqrt{8(1+\kappa^2)}\dot{\phi}^1_n - 2(2+\kappa^2)\phi^3_n = 0 \quad (5.163)$$
$$\ddot{\phi}^4_n + n^2\phi^4_n + 2in\phi^5_n + \phi^4_n = 0 \quad (5.164)$$
$$\ddot{\phi}^5_n + n^2\phi^5_n - 2in\phi^4_n + \phi^5_n = 0 \quad (5.165)$$
$$\ddot{\phi}^6_n + n^2\phi^6_n = 0 \quad (5.166)$$
$$\ddot{\phi}^7_n + n^2\phi^7_n = 0 \quad (5.167)$$
$$\ddot{\phi}^8_n + n^2\phi^8_n = 0 \quad (5.168)$$

The mode $n = 0$ corresponds to a σ independent perturbation, so it is not a true perturbation. It merely corresponds to changing the initial configuration, thus we take $|n| \geq 1$. It follows that the 3 last equations are solved by trigonometric functions, and are therefore stable. The first 3 equations are of course the same as (5.76) when written in first order form, thus they are stable for $\kappa^2 \leq 5/2$. We are then left with the last two equations which can be written (we define $p^4_n = \dot{\phi}^4_n$ and $p^5_n = \dot{\phi}^5_n$)

$$\begin{pmatrix} \dot{\phi}^4_n \\ \dot{p}^4_n \\ \dot{\phi}^5_n \\ \dot{p}^5_n \end{pmatrix} + \begin{pmatrix} 0 & -1 & 0 & 0 \\ 1+n^2 & 0 & 2in & 0 \\ 0 & 0 & 0 & -1 \\ -2in & 0 & 1+n^2 & 0 \end{pmatrix} \begin{pmatrix} \phi^4_n \\ p^4_n \\ \phi^5_n \\ p^5_n \end{pmatrix} = \begin{pmatrix} 0 \\ 0 \\ 0 \\ 0 \end{pmatrix} \quad (5.169)$$

To have stability we must have that all eigenvalues are imaginary, as follows from (5.80). The eigenvalues are

$$\lambda_+ = \pm i|1-n|$$
$$\lambda_- = \pm i|1+n| \quad (5.170)$$

thus we have stability. The modes corresponding to $n = \pm 1$ are zero-modes. In conclusion, we have when $A = 0$ stability for $\kappa^2 \leq 5/2$ corresponding to spin

$$S \leq \frac{5\sqrt{7}}{8\sqrt{2}H^2\alpha'} \quad (5.171)$$

Another special case of importance is $A = 1$, corresponding to a string which starts at one of the poles and approaches the equator for $\tau \to \infty$. Unfortunately, the equations do not really simplify in this limit. However, if we only look at the asymtotic equations, we get

$$\ddot{\phi}^1_n + n^2\phi^1_n - 2(\kappa in\phi^2_n + \sqrt{2}\kappa\dot{\phi}^3_n) = 0 \tag{5.172}$$
$$\ddot{\phi}^2_n + n^2\phi^2_n + 2(\kappa in\phi^1_n + \sqrt{2}\dot{\phi}^4_n) + 4(\kappa^2 - 1)\phi^2_n = 0 \tag{5.173}$$
$$\ddot{\phi}^3_n + n^2\phi^3_n + 2(in\phi^4_n + \sqrt{2}\kappa\dot{\phi}^1_n) - 2(\kappa^2 - 1)\phi^3_n = 0 \tag{5.174}$$
$$\ddot{\phi}^4_n + n^2\phi^4_n - 2(in\phi^3_n + \sqrt{2}\dot{\phi}^2_n) = 0 \tag{5.175}$$
$$\ddot{\phi}^5_n + (n^2 - 1)\phi^5_n = 0 \tag{5.176}$$
$$\ddot{\phi}^6_n + (n^2 - 1)\phi^6_n = 0 \tag{5.177}$$
$$\ddot{\phi}^7_n + (n^2 - 1)\phi^7_n = 0 \tag{5.178}$$
$$\ddot{\phi}^8_n + (n^2 - 1)\phi^8_n = 0 \tag{5.179}$$

The last four equations are solved by trigonometric functions except when $n = 0, \pm 1$, which are zero-modes, and therefore are stable.

Let us then look at the first 4 equations and apply the method of the previous case

$$\begin{pmatrix} \dot{\phi}^1_n \\ \dot{p}^1_n \\ \dot{\phi}^2_n \\ \dot{p}^2_n \\ \dot{\phi}^3_n \\ \dot{p}^3_n \\ \dot{\phi}^4_n \\ \dot{p}^4_n \end{pmatrix} + \mathcal{A} \begin{pmatrix} \phi^1_n \\ p^1_n \\ \phi^2_n \\ p^2_n \\ \phi^3_n \\ p^3_n \\ \phi^4_n \\ p^4_n \end{pmatrix} = \begin{pmatrix} 0 \\ 0 \\ 0 \\ 0 \\ 0 \\ 0 \\ 0 \\ 0 \end{pmatrix} \tag{5.180}$$

where the matrix $\mathcal{A}$ is

$$\begin{pmatrix} 0 & -1 & 0 & 0 & 0 & 0 & 0 & 0 \\ n^2 & 0 & -2\kappa in & 0 & 0 & -2^{3/2}\kappa & 0 & 0 \\ 0 & 0 & 0 & -1 & 0 & 0 & 0 & 0 \\ 2\kappa in & 0 & n^2 + 4(\kappa^2 - 1) & 0 & 0 & 0 & 0 & 2^{3/2} \\ 0 & 0 & 0 & 0 & 0 & -1 & 0 & 0 \\ 0 & 2^{3/2}\kappa & 0 & 0 & n^2 - 2(\kappa^2 - 1) & 0 & 2in & 0 \\ 0 & 0 & 0 & 0 & 0 & 0 & 0 & -1 \\ 0 & 0 & 0 & -2^{3/2} & -2in & 0 & n^2 & 0 \end{pmatrix} \tag{5.181}$$

For $\lambda^2 = l$ the characteristic polyomium is

$$\begin{aligned} p(l) = \; & l^4 + l^3(4n^2 + 10\kappa^2 + 6) + l^2(6n^4 + 24\kappa^4 + 32\kappa^2 + 18\kappa^2 n^2 + 6n^2 + 8) \\ & + l(4n^6 - 6n^4 + 16n^2 + 6n^4\kappa^2 + 24\kappa^4 n^2 + 24\kappa^2 n^2) \\ & + n^8 - 6n^6 + 8n^4 - 2n^6\kappa^2 + 8n^4\kappa^2 \end{aligned} \tag{5.182}$$

We first show that the 3 extrema fall at negative l, i.e. that the roots of

$$\begin{aligned} p'(l) = \quad & 4l^3 + 3l^2(4n^2 + 10\kappa^2 + 6) \\ & +2l(6n^4 + 24\kappa^4 + 32\kappa^2 + 18\kappa^2 n^2 + 6n^2 + 8) \\ & +(4n^6 - 6n^4 + 16n^2 + 6n^4\kappa^2 + 24\kappa^4 n^2 + 24\kappa^2 n^2) \end{aligned} \tag{5.183}$$

are negative. For that purpose, it is enough to show that the 2 extrema of $p'(0)$ fall at negative l, and that $p'(0) > 0$. We get

$$p''(l) = 12l^2 + 6l(4n^2 + 10\kappa^2 + 6) + 2(6n^4 + 24\kappa^4 + 32\kappa^2 + 18\kappa^2 n^2 + 6n^2 + 8) = 0 \tag{5.184}$$

It is easily seen that both of the solutions of this equation fulfill the requirement for arbitrary n, κ. We then have to look at $p'(0)$

$$p'(0) = 4n^6 - 6n^4 + 16n^2 + 6n^4\kappa^2 + 24\kappa^4 n^2 + 24\kappa^2 n^2 \tag{5.185}$$

Given that the last three terms are positive, and that the first three can be written as

$$4n^6 - 6n^4 + 16n^2 = n^2(2n^2 - 2)^2 + 2n^4 + 12n^2 \tag{5.186}$$

it is seen that $p'(0) > 0$. Now we just have to find the condition that $p(0) \geq 0$, which leads to

$$(4 - n^2)(2 + 2\kappa^2 - n^2) \geq 0 \tag{5.187}$$

As in the previous case, for $n = 0, 1, 2$, κ is arbitrary. For $n = 3, 4, 5, ...$ we get $\kappa \leq \sqrt{\frac{7}{2}}, \sqrt{7}, \sqrt{\frac{23}{2}}, ...$, respectively. Thus we end up with $\kappa^2 \leq 7/2$, corresponding to the spin

$$S = \frac{15}{8\sqrt{2}H^2\alpha'} \tag{5.188}$$

indicating that we have better stability as compared to the $A = 0$ case.

There is actually another case where we can solve the equations analytically, namely when $\kappa = A$. As seen from (3.7) and (3.9), this limit corresponds to zero spin, $S = 0$. We get the equations for $A \leq 1$

$$\ddot{\phi}^1_n + n^2\phi^1_n - \sqrt{4(1 + A^2)}\dot{\phi}^3_n - 2in\phi^2_n = 0 \tag{5.189}$$

$$\ddot{\phi}^2_n + n^2\phi^2_n + \sqrt{4(1 + A^2)}\dot{\phi}^4_n + 2in\phi^1_n = 0 \tag{5.190}$$

$$\ddot{\phi}^3_n + n^2\phi^3_n + \sqrt{4(1 + A^2)}\dot{\phi}^1_n + 2in\phi^4_n = 0 \tag{5.191}$$

$$\ddot{\phi}^4_n + n^2\phi^4_n - \sqrt{4(1 + A^2)}\dot{\phi}^2_n - 2in\phi^3_n = 0 \tag{5.192}$$

$$\ddot{\phi}^5_n + n^2\phi^5_n - \frac{2 - A^2\text{sn}^2(\tau|A^2))}{\text{sn}^2(\tau|A^2)}\phi^5_n = 0 \tag{5.193}$$

$$\ddot{\phi}^6_n + n^2\phi^6_n - A^2(2\text{sn}^2(\tau|A^2) - 1)\phi^6_n = 0 \tag{5.194}$$

$$\ddot{\phi}^7_n + n^2\phi^7_n - A^2(2\text{sn}^2(\tau|A^2) - 1)\phi^7_n = 0 \tag{5.195}$$

$$\ddot{\phi}^8_n + n^2\phi^8_n - A^2(2\text{sn}^2(\tau|A^2) - 1)\phi^8_n = 0 \tag{5.196}$$

and for $A \geq 1$

$$\ddot{\phi}_n^1 + n^2\phi_n^1 - \sqrt{4(1+A^2)}\dot{\phi}_n^3 - 2in\phi_n^2 = 0 \tag{5.197}$$
$$\ddot{\phi}_n^2 + n^2\phi_n^2 + \sqrt{4(1+A^2)}\dot{\phi}_n^4 + 2in\phi_n^1 = 0 \tag{5.198}$$
$$\ddot{\phi}_n^3 + n^2\phi_n^3 + \sqrt{4(1+A^2)}\dot{\phi}_n^1 + 2in\phi_n^4 = 0 \tag{5.199}$$
$$\ddot{\phi}_n^4 + n^2\phi_n^4 - \sqrt{4(1+A^2)}\dot{\phi}_n^2 - 2in\phi_n^3 = 0 \tag{5.200}$$
$$\ddot{\phi}_n^5 + n^2\phi_n^5 - \frac{A^2(2 - \text{sn}^2(A\tau|1/A^2))}{\text{sn}^2(A\tau|1/A^2)}\phi_n^5 = 0 \tag{5.201}$$
$$\ddot{\phi}_n^6 + n^2\phi_n^6 - (2\text{sn}^2(A\tau|1/A^2) - A^2)\phi_n^6 = 0 \tag{5.202}$$
$$\ddot{\phi}_n^7 + n^2\phi_n^7 - (2\text{sn}^2(A\tau|1/A^2) - A^2)\phi_n^7 = 0 \tag{5.203}$$
$$\ddot{\phi}_n^8 + n^2\phi_n^8 - (2\text{sn}^2(A\tau|1/A^2) - A^2)\phi_n^8 = 0 \tag{5.204}$$

The first four equations, which are identical in the two situations, can be shown to lead to oscillating solutions only (and zero-modes) by the same method as used before, so we skip the details. The last three equations in both situations can also not lead to divergenses, since the term in front of ϕ_n^i $(i = 6, 7, 8)$ is positive, except for $n = 0$ which is a zero-mode. Thus we are left with the equations for ϕ_n^5. These equations lead to instabilities since the term in front of ϕ_n^5 is periodically minus infinity.

It is not surprising that the pulsating strings become classically unstable for $S \approx 0$. Notice that the length of the unperturbed string is

$$L(\tau) = \int_0^{2\pi} \frac{ds}{d\sigma} d\sigma = \frac{2\pi}{H}\sqrt{H^2 r_0^2 + \sin^2\theta(\tau)} \tag{5.205}$$

$S \approx 0$ corresponds to $r_0 \approx 0$ and then we get $L(\tau) = 2\pi H^{-1}|\sin\theta(\tau)|$. It follows that the string oscillates between its maximal size and zero size. It is well-known that the classical perturbations blow up when a circular string collapses to a point. But of course the classical approximation cannot be trusted in this limit.

5.4 Numerical Analysis

Given the equations as they stand, eqs.(5.153)-(5.160) and their counterparts for $A \geq 1$, we now make some general comments, irrespective of the value of A. We have 5 coupled and 3 uncoupled second order ordinary differential equations for the Fourier components ϕ_n^i. Needless to say, except for the few special cases considered in the previous subsection, any attempts at solving the equations analytically end up being an exercise in futility. The correct way of going about the problem is to solve the equations numerically but, before we do so, let us first see how much information can be obtained from purely theoretical considerations.

Let us see what we can say about the uncoupled equations for general A. For $A \leq 1$ the last 3 equations were (we skip the i index)

$$\ddot{\phi}_n + \left(A^2 + n^2 - 2A^2\text{sn}^2(\tau|A^2)\right)\phi_n = 0 \tag{5.206}$$

which can be written as

$$\ddot{\phi}_n + c_1(\tau)\phi_n = 0 \tag{5.207}$$

and $c_1(\tau)$ can in turn be written as $c_1(\tau) = n^2 + A^2\text{cn}^2(\tau|A^2) - A^2\text{sn}(\tau|A^2)$, which is easily seen always to be positive, for $n \neq 0$. Notice that $n = 0$ is the zero-mode, which merely corresponds to changing the initial configuration. If, on the one hand, ϕ_n becomes large and positive in (5.207), its acceleration becomes negative, leading to a deceleration. If, on the other hand, ϕ_n becomes large and negative, the acceleration will be positive, and ϕ_n will start to become larger again, thus making ϕ_n, all in all, non-divergent.

For $A \geq 1$, we get

$$\ddot{\phi}_n + c_2(\tau)\phi_n = 0 \tag{5.208}$$

with $c_2(\tau) = n^2 - 2\text{sn}^2(A\tau|1/A^2) + A^2$, which is again seen to be positive, for $n \neq 0$. By the same argument as for $A \leq 1$, we conclude that ϕ_n is non-divergent.

Thus we are left with the first 5 equations. As mentioned in the beginning of this subsection, the right way to go about solving the coupled equations for arbitrary A, is by numerical analysis. In what follows, we first convert a second order ordinary differential equation to 2 first order ordinary differential equations, and show how to solve them, writing down an appropriate algorithm. We touch on the subjects of boundary conditions, step size and number of steps, and how to analyse the numerical output.

A second order ordinary differential equation of the form

$$\ddot{y} = g(\tau, y, \dot{y}) \tag{5.209}$$

can be converted to a system of 2 first order ordinary differential equations by the following change of variables

$$u = y \quad , \quad v = \dot{y} \tag{5.210}$$

This leads to the following 2 first order differential equations

$$\begin{aligned} \dot{u} = v = f(\tau, u, v) \\ \dot{v} = g(\tau, u, v) \end{aligned} \tag{5.211}$$

which can now be solved numerically, using an appropriate algorithm.

A good safe choice is the classic fourth order Runge Kutta method which, for step size h, uses the following linear combination of 4 function evaluations

$$\begin{aligned} k &= hf(\tau_i, y_i) \\ m &= hf(\tau_i + \frac{1}{2}h, y_i + \frac{1}{2}k) \\ j &= hf(\tau_i + \frac{1}{2}h, y_i + \frac{1}{2}m) \\ l &= hf(\tau_i + h, y_i + j), \end{aligned} \tag{5.212}$$

which, together with the recursion relation

$$y_{i+1} = y_i + \frac{1}{6}k + \frac{1}{3}m + \frac{1}{3}j + \frac{1}{6}l \tag{5.213}$$

makes up the method.

In order to run this algorithm, we must provide the step size h, and the total number of steps, N. In addition to that, we must also supply boundary conditions. The final data was analysed by printing the 10 graphs for the 5 complex solutions, $\phi_n^1 - \phi_n^5$.

All the aforementioned issues have been dealt with on a trial and error basis. The step size has been made gradually smaller, until there were no noticeable changes in the output, i.e., no changes on the fifth decimal place. In terms of N, we looked at the graphs and determined, that roughly after 100000 points, it was clear whether the functions diverged or not.

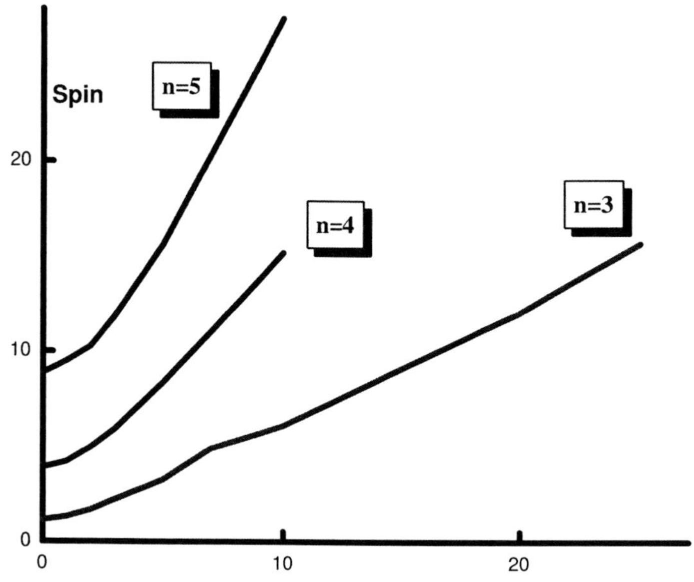

Figure 1: Critical spin in units of $(H^2\alpha')^{-1}$ as a function of A.

Ideally, the stability of the equations should be tested for an infinite number of possible boundary conditions, but since that is impossible, we have had to content ourselves with being able to only check a limited number of boundary conditions for the solutions. We have not made a major issue out of it, and have only varied the boundary conditions for a couple of stable and unstable solutions. This minor analysis showed, in terms of the behaviour of the functions for large τ, total insensitivity to varying the boundary conditions. Given that we are dealing with complex functions, we have chosen finite different complex boundary conditions around unity for all the functions. This means, that if we had a solution that were divergent for small τ, we would not see it in the solution graphs.

We now turn to the numerical results. For each (A, n) we integrate the 5 equations for $Hc_0 = \kappa > A$, to determine whether we have stability or not. Typically we have stability up to a critical κ, beyond which we have instability. The corresponding critical spin S, obtained from (3.9), is plotted in figure 1. for $A = 0, 1$ we found analytically that the modes $n = 0, 1, 2$ are either zero-modes or stable. This turns out to hold for arbitrary A. We also saw in the same cases that $n = 3$ was the most unstable mode. This also turns out to be the case for arbitrary A; see again figure 1.

More precisely, in figure 1 we show the maximal spin for which we have stability as a function of A. We only show the curves for the modes $n = 3, 4, 5$, just to show that $n = 3$ is the most unstable mode, as already mentioned.

The numerical results are of course in agreement with the analytical results for $A = 0, 1$, and from figure 1 we can then finally conclude that the inclusion of pulsation leads to better stability. In other words, the more pulsation we have (the higher A), the higher spin we can allow and still have stability.

We close this section with a few words about the zero-modes for $|n| = 0, 1, 2$. The zero-modes correspond to a change of the initial configuration. This could be a change of the radius of the circular string, or a change in the plane in which the unperturbed string is lying. These zero-modes are discussed in more detail in [50, 51, 52]. There is also a $|n| = 2$ zero-mode corresponding to the fact that the two angular momenta are identical [27].

6 Conclusion

We first reviewed most of the known string solitons in AdS_5. We then extended the two-spin string soliton to include pulsation in S^5. We went through the world-sheet covariant analysis of perturbations around string solitons in general, and took two examples: The two-spin string soliton and the pulsating two-spin string soliton. We finally showed that pulsation improves the stability properties of the two-spin string soliton.

A Appendix

We have the $AdS_5 \times S^5$ line-element

$$\begin{aligned} ds^2 &= -(1+H^2r^2)dt^2 + \frac{dr^2}{1+H^2r^2} + r^2(d\beta^2 + \sin^2\beta d\phi^2 + \cos^2\beta d\tilde{\phi}^2) \\ &+ \frac{1}{H^2}(d\theta^2 + \sin^2\theta d\psi^2 + \cos^2\theta(d\psi_1^2 + \sin^2\psi_1 d\psi_2^2 + \cos^2\psi_1 d\psi_3^2)) \end{aligned} \quad (A.1)$$

It is convenient to use the notation ($\mu = 0, 1, \ldots, 9$)

$$X^\mu = (t, r, \beta, \phi, \tilde{\phi}, \theta, \psi, \psi_1, \psi_2, \psi_3) \quad (A.2)$$

Then the metric is

$$\begin{aligned}
G_{tt} &= -(1+H^2r^2), \quad G_{rr} = (1+H^2r^2)^{-1}, \\
G_{\beta\beta} &= r^2, \quad G_{\phi\phi} = r^2 \sin^2\beta \\
G_{\tilde{\phi}\tilde{\phi}} &= r^2\cos^2\beta, \quad G_{\theta\theta} = H^{-2}, \\
G_{\psi\psi} &= H^{-2}\sin^2\theta, \quad G_{\psi_1\psi_1} = H^{-2}\cos^2\theta, \\
G_{\psi_2\psi_2} &= H^{-2}\cos^2\theta\sin^2\psi_1, \quad G_{\psi_3\psi_3} = H^{-2}\cos^2\theta\cos^2\psi_1
\end{aligned} \tag{A.3}$$

The Christoffel symbols are

$$\Gamma^{\mu}_{\rho\sigma} = \frac{1}{2}G^{\mu\nu}(G_{\nu\rho,\sigma} + G_{\nu\sigma,\rho} - G_{\rho\sigma,\nu}) \tag{A.4}$$

which leads to

$$\begin{array}{lll}
\Gamma^t_{rt} = \frac{H^2r}{1+H^2r^2} & \Gamma^r_{rr} = \frac{-H^2r}{1+H^2r^2} & \Gamma^r_{tt} = (1+H^2r^2)H^2r \\
\Gamma^r_{\beta\beta} = -(1+H^2r^2)r & \Gamma^r_{\phi\phi} = -(1+H^2r^2)r\sin^2\beta & \Gamma^r_{\tilde{\phi}\tilde{\phi}} = -(1+H^2r^2)r\cos^2\beta \\
\Gamma^\beta_{\beta r} = \frac{1}{r} & \Gamma^\beta_{\phi\phi} = -\cos\beta\sin\beta & \Gamma^\beta_{\tilde{\phi}\tilde{\phi}} = \cos\beta\sin\beta \\
\Gamma^\phi_{\phi r} = \frac{1}{r} & \Gamma^\phi_{\phi\beta} = \frac{\cos\beta}{\sin\beta} & \Gamma^{\tilde{\phi}}_{\tilde{\phi}r} = \frac{1}{r} \\
\Gamma^{\tilde{\phi}}_{\tilde{\phi}\beta} = \frac{-\sin\beta}{\cos\beta} & \Gamma^\theta_{\psi\psi} = -\cos\theta\sin\theta & \Gamma^\theta_{\psi_1\psi_1} = \cos\theta\sin\theta \\
\Gamma^\theta_{\psi_2\psi_2} = \sin^2\psi_1\cos\theta\sin\theta & \Gamma^\theta_{\psi_3\psi_3} = \cos^2\psi_1\cos\theta\sin\theta & \Gamma^\psi_{\psi\theta} = \frac{\cos\theta}{\sin\theta} \\
\Gamma^{\psi_1}_{\psi_1\theta} = \frac{-\sin\theta}{\cos\theta} & \Gamma^{\psi_1}_{\psi_2\psi_2} = -\cos\psi_1\sin\psi_1 & \Gamma^{\psi_1}_{\psi_3\psi_3} = \cos\psi_1\sin\psi_1 \\
\Gamma^{\psi_2}_{\psi_2\theta} = \frac{-\sin\theta}{\cos\theta} & \Gamma^{\psi_2}_{\psi_2\psi_1} = \frac{\cos\psi_1}{\sin\psi_1} & \Gamma^{\psi_3}_{\psi_3\theta} = \frac{-\sin\theta}{\cos\theta} \\
\Gamma^{\psi_3}_{\psi_3\psi_1} = \frac{-\sin\psi_1}{\cos\psi_1} & &
\end{array}$$

The Riemann tensor splits into two parts

$$\begin{aligned}
R^{AdS}_{\mu\nu\rho\sigma} &= -H^2(G^{AdS}_{\mu\rho}G^{AdS}_{\nu\sigma} - G^{AdS}_{\nu\rho}G^{AdS}_{\mu\sigma}); \; (\mu,\nu,\rho,\sigma) \in \{0,1,2,3,4\} \\
R^{S}_{\mu\nu\rho\sigma} &= H^2(G^{S}_{\mu\rho}G^{S}_{\nu\sigma} - G^{S}_{\nu\rho}G^{S}_{\mu\sigma}); \; (\mu,\nu,\rho,\sigma) \in \{5,6,7,8,9\}
\end{aligned} \tag{A.5}$$

Acknowledgements

We would like to thank N. Sibani for secretarial help, and L. Glazov for help with the numerical work.

References

[1] J. Maldacena, *Adv. Math. Phys.* **2**, 231 (1998), *Int. J. Theor. Phys.* **38**, 1113 (1999).

[2] S.S. Gubser, I.R. Klebanov and A.M. Polyakov, *Phys. Lett.* **B428**, 105 (1998).

[3] E. Witten, *Adv. Theor. Math. Phys.* **2**, 253 (1998).

[4] H.J. de Vega and I.L. Egusquiza, *Phys. Rev.* **D54**, 7513 (1996).

[5] H.J. de Vega, A.L. Larsen and N. Sánchez, *Phys. Rev.* **D51**, 6917 (1995).

[6] S.S. Gubser, I.R. Klebanov and A.M. Polyakov, *Nucl. Phys.* **B636**, 99 (2002).

[7] D.J. Gross and F. Wilczek, *Phys. Rev.* **D9**, 980 (1974).

[8] H. Georgi and D. Politzer, *Phys. Rev.* **D9**, 416 (1974).

[9] E.G. Floratos, D.A. Ross and C.T. Sachrajda, *Nucl. Phys.* **B129**, 66 (1977).

[10] G.P. Korchemsky, *Mod. Phys. Lett.* **A4**, 1257 (1989).

[11] F.A. Dolan and H. Osborn, *Nucl. Phys.* **B629**, 3 (2002).

[12] S. Frolov and A.A. Tseytlin, *JHEP* **0206**, 007 (2002).

[13] A.A. Tseytlin, *Int. J. Mod. Phys.* **A18**, 981 (2003).

[14] J.G. Russo, *JHEP* **0206**, 038 (2002).

[15] A. Armoni, J.L.F. Barbon and A.C. Petkou, *JHEP* **0206**, 058 (2002).

[16] G. Mandal, *Phys. Lett.* **B543**, 81 (2002).

[17] J.A. Minahan, *Nucl. Phys.* **B648**, 203 (2003).

[18] A. Armoni, J.L.F. Barbon and A.C. Petkou, *JHEP* **0210**, 069 (2002).

[19] M. Axenides and E. Floratos, *Nucl. Phys.* **B662**, 170 (2003).

[20] M. Alishahiha and A.E. Mosaffa, *JHEP* **0210**, 060 (2002).

[21] A. Buchel, *Phys. Rev.* **D67**, 066004 (2003).

[22] R.C. Rashkov and K.S. Viswanathan, *"Rotating Strings with B-Field"*, *hep-th/0211197* (unpublished).

[23] S. Ryang, *JHEP* **0304**, 045 (2003).

[24] D. Aleksandrova and P. Bozhilov, *JHEP* **0308**, 018 (2003).

[25] D. Aleksandrova and P. Bozhilov, *"On The Classical String Solutions and String/Field Theory Duality 2"*, *hep-th/0308087* (unpublished).

[26] A. Khan and A.L. Larsen, *Phys. Rev.* **D69** 026001 (2004).

[27] S. Frolov and A.A. Tseytlin, *Nucl. Phys.* **B668**, 77 (2003).

[28] J. Engquist, J.A. Minahan and K. Zarembo, *JHEP* **0311**, 063 (2003).

[29] N. Beisert, J.A. Minahan, M. Staudacher and K. Zarembo, *JHEP* **0309**, 010 (2003).

[30] S. Frolov and A.A. Tseytlin, *JHEP* **0307**, 016 (2003).

[31] S. Frolov and A.A. Tseytlin, *Phys. Lett.* **B570**, 96 (2003).

[32] G. Arutyunov, S. Frolov, J. Russo and A.A. Tseytlin, *Nucl. Phys.* **B671**, 3 (2003).

[33] N. Beisert, S. Frolov, M. Staudacher and A.A. Tseytlin, *JHEP* **0310**, 037 (2003).

[34] G. Arutyunov, J. Russo and A.A. Tseytlin, *Phys. Rev.* **D69**, 086009 (2004).

[35] B. Stefanski, *JHEP* **0403**, 057 (2004).

[36] N. Kim, *Phys. Rev.* **D69**, 126002 (2004).

[37] A.A. Tseytlin, *"Spinning Strings and AdS/CFT Duality"*, *hep-th/0311139* (unpublished).

[38] S. Kar and S. Mahapatra, *Class. Quant. Grav.* **15**, 1421 (1998).

[39] V. Frolov, S. Hendy and J.P. De Villiers, *Class. Quant. Grav.* **14**, 1099 (1997).

[40] M. Abramowitz and I.A. Stegun, *"Handbook of Mathematical Functions"*, Dover Publications (1972).

[41] A.L. Larsen and A. Khan, *Nucl. Phys.* **B686** 75 (2004).

[42] H.J. de Vega and N. Sánchez, *Nucl. Phys.* **B317**, 706 (1989).

[43] H.J. de Vega and N. Sánchez, *Nucl. Phys.* **B317**, 731 (1989).

[44] H.J. de Vega and N. Sánchez, *Phys. Lett.* **B244**, 215 (1990).

[45] A.L. Larsen and V.P. Frolov, *Nucl. Phys.* **B414**, 129 (1994).

[46] J. Guven, *Phys. Rev.* **D48**, 4604 (1993).

[47] J. Guven, *Phys. Rev.* **D48**, 5562 (1993).

[48] S. Kar, *Phys. Rev.* **D54**, 6408 (1996).

[49] A.L. Larsen, *Phys. Rev.* **D50** 2623 (1994).

[50] J. Garriga and A. Vilenkin, *Phys. Rev.* **D44** 1007 (1991).

[51] J. Garriga and A. Vilenkin, *Phys. Rev.* **D45** 3469 (1992).

[52] J. Garriga and A. Vilenkin, *Phys. Rev.* **D47** 3265 (1993).

In: Advances in Soliton Research
Editor: L.V. Chen, pp. 45-103

ISBN 1-59454-769-6

Chapter 2

SOLITON SELF-MODULATION OF THE TURBULENCE AMPLITUDE AND PLASMA ROTATION

Florin Spineanu*and Madalina Vlad†
Association Euratom-C.E.A. sur la Fusion, C.E.A.-Cadarache,
F-13108 Saint-Paul-lez-Durance, France
Association Euratom-MEC Romania,
National Institute for Laser, Plasma and Radiation Physics,
P.O.Box MG-36, Magurele, Bucharest, Romania

Abstract

The space-uniform amplitude envelope of the Ion Temperature Gradient driven turbulence in Tokamak is unstable to small perturbations and evolves to nonuniform, soliton-like modulated profiles. The induced poloidal asymmetry of the transport fluxes can generate spontaneous poloidal spin-up of the tokamak plasma.

1 Introduction

The ion temperature gradient turbulence, considered a major source of anomalous transport in the tokamak plasma, is characterized by the coexistence of irregular patterns (randomly fluctuating field) and intermittent robust cuasi-coherent structures. In closely related fluid models (for exemple, in the physics of atmosphere) modulational instabilities are known to produce solitary structures on the envelope of the fluctuating field. In the case of the tokamak, this can be particularly important since a poloidally nonuniform amplitude of the turbulence generates nonuniform transport rates. For a sufficiently high nonuniformity the

*E-mail address: florin@drfc.cad.cea.fr
†E-mail address: madi@drfc.cad.cea.fr

torque arising via the mechanism initially mentioned by Stringer [2] may overcome the rotation damping due to the poloidal magnetic pumping. We have proved that this is indeed possible in Ref.[1]. In this Chapter we extend this argument and prove that the uniform poloidal envelope of the ITG turbulence is an unstable state. It appears that the ITG turbulence has intrinsic resources to generate poloidal rotation via a combination of mechanisms which is not directly related to the Reynolds stress, inverse cascade, direct-ion loss, or other classical sources of rotation. We provide an essentially mathematical explanation of the instability of the solution consisting of poloidally uniform envelope of the turbulence. Based on the geometrico-algebraic method of solving integrable nonlinear differential equations on periodic domains we invoke an existing result, that any perturbation removes the degeneracies due to coincident eigenvalues in the main spectrum of the Lax operator, thus changing the topology of the hyperelliptic Riemann surface that provides the solution. The perturbed solution separates exponentially (in function space) from the initial one (uniform envelope) and this yields an exponential growth of the poloidal nonuniformity. The magnitude of the torque may be comparable to the poloidal magnetic damping.

There are many works in plasma physics related to the soliton dynamics [3], [4], [5], [6], [8], etc. In a recent paper the focusing solution has been used to study the formation of coherent motion or intermittent patterns (*streamers*) [3]. Our work is basically a mathematical approach, using well developed technics related to the Inverse Scattering Transform method for periodic domains. However, more physical analysis may become possible combining the knowledge of the spectral properties of the ion turbulence with the nonlinear stability properties.

This Chapter consists of several parts that may appear as developing separately: the derivation of the ion equation (barotropic equation), the multiple space-time scales analysis, the solution of the Nonlinear Schrodinger Equation and the stability of the solution, the torque arising from the Stringer mechanism and the possibility of the rotation. For being self-contained the Chapter also includes a review of the multiple space-time analysis (from the atmospheric physics applications) and of the geometric-algebraic method of integration of the Nonlinear Schrodinger Equation. Even if these parts could be found in basic works (see the references) they are included both for clarity and for their extreme importance for further applications related to our problem or independent of this.

2 The Slab Model of the Ion Mode Instability

We consider cylindrical geometry with circular magnetic surfaces. Locally the model can be reduced to a slab geometry with (x, y) cartezian coordinates replacing respectively the radius and poloidal angle coordinates (r, θ). At equilibrium the plasma parameters are constant on the magnetic surfaces. The effects of the toroidicity and of the particle drifts are not included instead the nonlinearity related to the ion polarization drift is fully retained. The plasma model consists of the continuity equations and the equations of motion for electrons and ions:

$$\frac{\partial n_\alpha}{\partial t} + \nabla \cdot (n_\alpha \mathbf{v}_\alpha) = 0 \qquad (1)$$

$$m_\alpha n_\alpha \left(\frac{\partial \mathbf{v}_\alpha}{\partial t} + (\mathbf{v}_\alpha \cdot \nabla) \mathbf{v}_\alpha \right) = -\nabla p_\alpha - \nabla \cdot \pi_\alpha + e_\alpha n_\alpha (\mathbf{E} + \mathbf{v}_\alpha \times \mathbf{B}) + \mathbf{R}_\alpha \quad (2)$$

where $\alpha = e, i$. The friction forces $\mathbf{R}_e = -\mathbf{R}_i = -n|e| \mathbf{J}_\| / \sigma_\|$, which are important for the parallel electron momentum balance, vanish for infinite plasma conductivity, which we will assume. The collisional viscosity $\pi_{e,i}$ will be neglected as well. However we will need to include it later when we will consider the balance of the forces contributing to the poloidal rotation. The electron and ion temperatures are considered constant along the magnetic lines $\nabla_\| T_{e,i} = 0$. The equilibrium quantities are perturbed by the wave potential ϕ: $n = n_0 + \widetilde{n}$. A sheared poloidal plasma rotation is included, and we later will make explicit the corresponding part in the potential, ϕ_0.

The momentum conservation equations are used to determine the perpendicular velocities of the electrons and ions. The parallel momentum conservation equation for electrons, in the absence of *dissipation* or *drifts* gives the adiabatic distribution of the density fluctuation. The velocities are introduced in the continuity equations to find the dynamical equations for the density and electric potential.

From the equations of motion for the *ions* the velocities are obtained in the form:

$$\mathbf{v}_i = \mathbf{v}_{\perp i} = \mathbf{v}_{dia,i} + \mathbf{v}_E + \mathbf{v}_{pol,i} \quad (3)$$

where the ion diamagnetic velocity is

$$\mathbf{v}_{dia,i} = \frac{1}{n_i} \frac{1}{m_i \Omega_i} \widehat{\mathbf{n}} \times \nabla p_i \quad (4)$$

The versor of the magnetic field is $\widehat{\mathbf{n}}$; the versors along the transversal coordinate axis (x, y) will be noted $(\widehat{\mathbf{e}}_x, \widehat{\mathbf{e}}_y)$. The **ion-polarization velocity** is:

$$\begin{aligned} \mathbf{v}_{pol,i} &= -\frac{1}{|e| n_i B} n_i m_i \left(\frac{\partial \mathbf{v}_E}{\partial t} + (\mathbf{v}_E \cdot \nabla) \mathbf{v}_E \right) \times \widehat{\mathbf{n}} \\ &= -\frac{1}{B \Omega_i} \left(\frac{\partial}{\partial t} + (\mathbf{v}_E \cdot \nabla_\perp) \right) \nabla_\perp \phi \end{aligned} \quad (5)$$

Using the notation $\frac{d}{dt} \equiv \frac{\partial}{\partial t} + (\mathbf{v}_E \cdot \nabla_\perp)$ the perpendicular ion velocity can be written

$$\mathbf{v}_{\perp,i} = \frac{T_i}{|e| B} \frac{1}{n_i} \frac{dn_i}{dr} \widehat{\mathbf{e}}_y - \frac{1}{B \Omega_i} \frac{d}{dt} (\nabla_\perp \phi) + \frac{-\nabla \phi \times \widehat{\mathbf{n}}}{B} \quad (6)$$

The equation for the velocity of the *electrons* is

$$\mathbf{v}_e = \mathbf{v}_{\perp,e} + \mathbf{v}_{\|,e} \quad (7)$$

$$\begin{aligned} \mathbf{v}_{\perp e} &= \mathbf{v}_E + \mathbf{v}_{dia,e} \\ &= \frac{-\nabla \phi \times \widehat{\mathbf{n}}}{B} + \frac{1}{n} \frac{1}{m_e \Omega_e} (\widehat{\mathbf{n}} \times \nabla p) \\ &= \frac{-\nabla \phi \times \widehat{\mathbf{n}}}{B} + \frac{T_e}{-|e| B} \frac{1}{n_0} \frac{dn_0}{dr} \widehat{\mathbf{e}}_y \end{aligned} \quad (8)$$

We assume neutrality $n_e = n_i = n$ and introduce the expressions of the velocities in the continuity equations for ions and for electrons.

The ion continuity equation is

$$\frac{\partial n}{\partial t} + \nabla \cdot (n\mathbf{v}_{\perp,i}) = 0 \tag{9}$$

The electron continuity equation is

$$\frac{\partial n}{\partial t} + \nabla_{\perp} \cdot (n\mathbf{v}_{\perp,e}) + \nabla_{\|} \left(n v_{\|e}\right) = 0 \tag{10}$$

The equations are substracted , to obtain

$$\begin{aligned} & -\nabla_{\|} \left(n v_{\|e}\right) \\ & -\frac{1}{B\Omega_i}\nabla_{\perp} \cdot \left(n\frac{d}{dt}\nabla_{\perp}\phi\right) \\ & +\nabla_{\perp} \cdot (n\mathbf{v}_{dia,i}) - \nabla_{\perp} \cdot (n\mathbf{v}_{dia,e}) \\ = \ & 0 \end{aligned} \tag{11}$$

From the last term in the left we get:

$$\begin{aligned} \frac{dn_0}{dx}\widehat{\mathbf{e}}_x \cdot \mathbf{v}_{dia,i} + (\nabla_{\perp}\widetilde{n}) \cdot \mathbf{v}_{dia,i} & = \\ & = -\frac{T_i}{T_e}\left(\frac{T_e}{|e|\,B}\frac{1}{n_0}\frac{dn_0}{dx}\right)\frac{\partial \widetilde{n}}{\partial y} \end{aligned} \tag{12}$$

Including the similar term for the electrons, we obtain

$$\begin{aligned} & -\nabla_{\|} \left(n v_{\|e}\right) \\ & -\frac{1}{B\Omega_i}\nabla_{\perp} \cdot \left(n\frac{d}{dt}\nabla_{\perp}\phi\right) \\ & -\frac{T_i}{T_e}\left(\frac{T_e}{|e|\,B}\frac{1}{n_0}\frac{dn_0}{dx}\right)\frac{\partial \widetilde{n}}{\partial y} - \left(\frac{T_e}{|e|\,B}\frac{1}{n_0}\frac{dn_0}{dx}\right)\frac{\partial \widetilde{n}}{\partial y} \\ = \ & 0 \end{aligned} \tag{13}$$

or:

$$-\left(1 + \frac{T_i}{T_e}\right)\left(\frac{T_e}{|e|\,B}\frac{1}{n_0}\frac{dn_0}{dx}\right)\frac{\partial \widetilde{n}}{\partial y} + \frac{1}{B\Omega_i}\nabla_{\perp} \cdot \left(n\frac{d}{dt}\nabla_{\perp}\phi\right) + \nabla_{\|}\left(n v_{\|e}\right) = 0 \tag{14}$$

From the continuity equation for electrons

$$\frac{\partial n}{\partial t} + \nabla_{\perp} \cdot (n\mathbf{v}_{\perp}) + \nabla_{\|} \cdot \left(n\mathbf{v}_{\|}\right) = 0 \tag{15}$$

we obtain

$$\begin{aligned} & \frac{\partial \widetilde{n}}{\partial t}+\frac{1}{B}\frac{\partial \phi}{\partial y}\frac{dn_0}{dx}++\mathbf{v}_{dia,e}\cdot\nabla_\perp \widetilde{n}+V_0\frac{\partial \widetilde{n}}{\partial y}+ \\ & +n_0\left(\nabla_\| \cdot \mathbf{v}_\|\right)+v_\| \nabla_\| \widetilde{n} \\ = \; & 0 \end{aligned} \tag{16}$$

where the seed poloidal velocity is $\mathbf{V}_0(x)=-\frac{d\phi_0(x)}{Bdx}\widehat{\mathbf{e}}_y$. The parallel momenttum balance gives the parallel electron velocity

$$\mathbf{v}_{\| e}=-\frac{\sigma_\|}{e^2}\nabla_\|\left(|e|\,\phi+T_e\ln\left(n/n_0\right)\right) \tag{17}$$

In the absence of friction $\left(\sigma_\| \to \infty\right)$ and of particle drifts the electron response is adiabatic

$$\frac{\widetilde{n}}{n_0}=-\frac{|e|\,\phi}{T_e} \tag{18}$$

and the potential is determined from Eq.(14). To develop separately the ion-polarization drift term, we introduce the notations:

$$\begin{aligned} W &\equiv \frac{1}{B\Omega_i}\nabla_\perp\cdot\left[\left(n_0+\widetilde{n}\right)\left(\frac{\partial}{\partial t}+\mathbf{v}_E\cdot\nabla_\perp\right)\nabla_\perp\phi\right] \\ &= \frac{1}{B\Omega_i}\nabla_\perp\cdot\left[\left(n_0+\widetilde{n}\right)\mathbf{I}\right] \end{aligned} \tag{19}$$

where we make explicit the electric potential ϕ_0 associated to the initial plasma poloidal rotation, $V_0\widehat{\mathbf{e}}_y$, and

$$\begin{aligned} \frac{1}{B}I_x &= \frac{\partial V_0}{\partial t}+\widetilde{v}_y\frac{\partial V_0}{\partial y}+V_0\frac{\partial V_0}{\partial y} \\ &+\frac{\partial \widetilde{v}_y}{\partial t}+\widetilde{v}_x\frac{\partial V_0}{\partial x}+V_0\frac{\partial \widetilde{v}_y}{\partial y} \\ &+\widetilde{v}_x\frac{\partial \widetilde{v}_y}{\partial x}+\widetilde{v}_y\frac{\partial \widetilde{v}_y}{\partial y}, \end{aligned} \tag{20}$$

$$\begin{aligned} \frac{1}{B}I_y &= -\frac{\partial \widetilde{v}_x}{\partial t}-V_0\frac{\partial \widetilde{v}_x}{\partial y} \\ &-\widetilde{v}_x\frac{\partial \widetilde{v}_x}{\partial x}-\widetilde{v}_y\frac{\partial \widetilde{v}_x}{\partial y} \end{aligned} \tag{21}$$

After simple calculations we obtain

$$\begin{aligned} W = \; & \frac{1}{B\Omega_i}\left(\frac{dn_0}{dx}\right)I_x+\frac{1}{B\Omega_i}n_0\left(\frac{\partial I_x}{\partial x}\right)+ \\ & +\frac{1}{B\Omega_i}\left(\frac{\partial \widetilde{n}}{\partial x}\right)I_x+\frac{1}{B\Omega_i}\widetilde{n}\left(\frac{\partial I_x}{\partial x}\right)+ \\ & +\frac{1}{B\Omega_i}n_0\left(\frac{\partial I_y}{\partial y}\right)+\frac{1}{B\Omega_i}\left(\frac{\partial \widetilde{n}}{\partial y}\right)I_y+\frac{1}{B\Omega_i}\widetilde{n}\left(\frac{\partial I_y}{\partial y}\right) \end{aligned} \tag{22}$$

2.1 The Mode Evolution in a Fixed Plasma Rotation Profile

We will assume that the mode evolves initially without perturbing the equilibrium profiles, in particular the seed poloidal rotation. This allows us to simplify the expressions above, taking:

$$\frac{\partial V_0}{\partial y}=0,\ \frac{\partial V_0}{\partial t}=0 \tag{23}$$

Collecting all what we have at this moment the ion continuity equation (14) becomes:

$$\begin{aligned}
&\nabla_{\parallel}\left(nv_{\parallel e}\right)+ \\
&-\left(1+\frac{T_i}{T_e}\right)\left(\frac{T_e}{|e|\,B}\frac{1}{n_0}\frac{dn_0}{dx}\right)\frac{\partial\widetilde{n}}{\partial y}+ \\
&+\frac{1}{B\Omega_i}n_0\left\{\left(\frac{\partial}{\partial t}+V_0\frac{\partial}{\partial y}\right)\nabla_{\perp}^2\widetilde{\phi}-V_0''\frac{\partial\widetilde{\phi}}{\partial y}+\left(\frac{-\nabla_{\perp}\phi\times\widehat{\mathbf{n}}}{B}\cdot\nabla_{\perp}\right)\nabla_{\perp}^2\widetilde{\phi}\right\} \\
&+\frac{1}{B\Omega_i}\left(\frac{dn_0}{dx}\right)\left[\left(\frac{\partial}{\partial t}+V_0\frac{\partial}{\partial y}\right)\widetilde{v}_y+\widetilde{v}_x\frac{\partial\widetilde{v}_y}{\partial x}+\widetilde{v}_y\frac{\partial\widetilde{v}_y}{\partial y}+\widetilde{v}_x\frac{\partial V_0}{\partial x}\right] \\
&+\widetilde{W}+\{\text{terms of the first line in the expressions of } I_x \text{ and derivatives}\}=0
\end{aligned} \tag{24}$$

where

$$\widetilde{W}=\frac{1}{B\Omega_i}\left(\frac{\partial\widetilde{n}}{\partial x}\right)I_x+\frac{1}{B\Omega_i}\widetilde{n}\left(\frac{\partial I_x}{\partial x}\right)+\frac{1}{B\Omega_i}\left(\frac{\partial\widetilde{n}}{\partial y}\right)I_y+\frac{1}{B\Omega_i}\widetilde{n}\left(\frac{\partial I_y}{\partial y}\right) \tag{25}$$

In Eq.(24) (which is exact) we shall make the following approximations:

- neglect the term containing the parallel electron velocity, assuming infinite electric conductivity;
- neglect the term which contains dn_0/dx since it is in the ratio $k:1/L_n$ with the other terms, and we consider
$$kL_n \ll 1 \tag{26}$$
- neglect $\widetilde{W}$; these terms are in the ratio $\widetilde{n}/n_0$ with the terms which are retained;
- neglect of the terms in the first lines of the expressions for I_x and of its derivatives. (These are the terms in the curly brakets, the last line). As explained above, we assume that the mode evolves in a background of fixed rotation profile, $V_0\left(x\right)$.

The resulting equation is

$$\begin{aligned}
&-B\Omega_i\left(1+\frac{T_i}{T_e}\right)\left(\frac{T_e}{|e|\,B}\frac{1}{n_0}\frac{dn_0}{dx}\right)\frac{\partial}{\partial y}\left(\frac{\widetilde{n}}{n_0}\right) \\
&+\left(\frac{\partial}{\partial t}+V_0\frac{\partial}{\partial y}\right)\nabla_{\perp}^2\widetilde{\phi}-V_0''\frac{\partial\widetilde{\phi}}{\partial y}+\left(\frac{-\nabla_{\perp}\phi\times\widehat{\mathbf{n}}}{B}\cdot\nabla_{\perp}\right)\nabla_{\perp}^2\widetilde{\phi}=0
\end{aligned} \tag{27}$$

and, replace the adiabatic form of the density perturbation $\widetilde{n}/n_0 = -|e|\widetilde{\phi}/T_e$. We define

$$\beta \equiv B\Omega_i \left(1+\frac{T_i}{T_e}\right)\left(\frac{T_e}{|e|\,B}\frac{1}{n_0}\frac{dn_0}{dx}\right)\frac{|e|}{T_e} = \Omega_i \left(1+\frac{T_i}{T_e}\right)\frac{1}{L_n} \tag{28}$$

and obtain

$$\beta\frac{\partial\widetilde{\phi}}{\partial y} + \left(\frac{\partial}{\partial t} + V_0\frac{\partial}{\partial y}\right)\nabla_\perp^2\widetilde{\phi} - V_0''\frac{\partial\widetilde{\phi}}{\partial y} + \left(\frac{-\nabla_\perp\phi\times\widehat{\mathbf{n}}}{B}\cdot\nabla_\perp\right)\nabla_\perp^2\widetilde{\phi} = 0 \tag{29}$$

which is the barotropic equation.

2.2 Nondimensional Form of the Equation

We consider that the ion mode propagates in the spatial (y), poloidal direction, over a length L. A typical value for the sheared poloidal rotation is noted U_0. We make the replacements

$$y \to Ly \tag{30}$$

$$t \to t\frac{L}{U_0} \tag{31}$$

$$\widetilde{\phi} \to \phi\frac{T_e}{|e|} \tag{32}$$

$$V_0 \to UU_0 \tag{33}$$

such that from now on y, t, ϕ and U are nondimensional quantities. We also change the radial coordinate into a dimensionless variable

$$x \to Lx \tag{34}$$

and rewrite the equation

$$\begin{aligned}&\left(\Omega_i\left(1+\frac{T_i}{T_e}\right)\frac{1}{L_n}\frac{L^2}{U_0}\right)\frac{\partial\phi}{\partial y} + \\ &+\left(\frac{\partial}{\partial t}+U\frac{\partial}{\partial y}\right)\nabla_\perp^2\phi \\ &-\frac{d^2U}{dx^2}\frac{\partial\phi}{\partial y} + \\ &+\left(\frac{1}{U_0}\frac{T_e}{|e|}\frac{1}{B}\frac{1}{L}\right)\left[\left(-\nabla_\perp\phi\times\widehat{\mathbf{n}}\right)\cdot\nabla_\perp\right]\nabla_\perp^2\phi = 0\end{aligned} \tag{35}$$

The coefficients are

$$\beta' \equiv \Omega_i\left(1+\frac{T_i}{T_e}\right)\frac{1}{L_n}\frac{L^2}{U_0} \quad \text{non-dimensional} \tag{36}$$

$$\varepsilon \equiv \frac{1}{U_0} \frac{T_e}{|e|} \frac{1}{B} \frac{1}{L} \text{ non-dimensional} \tag{37}$$

For an order of magnitude, ε is the ratio of the diamagnetic electron velocity to the rotation velocity U_0 multiplied by the ratio of the density gradient length to the length of the spatial domain. This quantity, ε is in general smaller than unity.

The quantity β' is the ratio of the ion cyclotron frequency to the inverse of the time required to cross the spatial domain with the typical flow velocity. Since the later (U_0/L) involves macroscopic quantities this ratio can be large. It is multiplied by the ratio of the spatial length to the density gradient length (these quantities can be comparable and the ratio not too different of unity).

We change the notations eliminating the primes. The equation becomes

$$\beta \frac{\partial \phi}{\partial y} + \left(\frac{\partial}{\partial t} + U \frac{\partial}{\partial y} \right) \nabla_\perp^2 \phi - \frac{d^2 U}{dx^2} \frac{\partial \phi}{\partial y} + \varepsilon \left[\left(-\nabla_\perp \phi \times \widehat{\mathbf{n}} \right) \cdot \nabla_\perp \right] \nabla_\perp^2 \phi = 0 \tag{38}$$

This is the *barotropic* equation, known in the physics of the atmosphere.

3 Multiple Time and Space Scale Analysis of the Ion Mode Equation Reduced to the Barotropic Form

The quasigeostrophic barotropic atmospheric model leads to the **barotropic equation** (see Horton [8] where the quasigeostrophic approximation is explained in relation with the Ertel's theorem). This equation has been studied in the atmosphere science by means of the *multiple space and time scales method.* Our aim is similar, *i.e.* to obtain an equation for the envelope of the fluctuating field ϕ in order to study the possible poloidally nonuniform profiles of the turbulence averaged amplitude. The connection between the original equation and the final slow-time and large-spatial scales equations is not simple : much numerical work is required in order to connect the input physical parameters of the plasma with the formal coefficients in the final equations. The detailed analytical work is presented in Ref([9]) and the necessary formulas are reproduced here for convenience. To simplify the expressions the following notation is introduced $\left[\phi, \nabla^2 \phi\right] \equiv \left[\left(-\nabla_\perp \phi \times \widehat{\mathbf{n}} \right) \cdot \nabla_\perp \right] \nabla_\perp^2 \phi$ and the equation can be written

$$\left(\frac{\partial}{\partial t} + U(x) \frac{\partial}{\partial y} \right) \nabla^2 \phi + \left(\beta - \frac{d^2 U(x)}{dx^2} \right) \frac{\partial \phi}{\partial y} = -\varepsilon \left[\phi, \nabla^2 \phi \right] \tag{39}$$

The analysis on multiple scales starts by introducing the variables on scales separated by the parameter ε:

$$T_1 = \varepsilon t, \quad T_2 = \varepsilon^2 t \tag{40}$$

$$Y_1 = \varepsilon y, \quad Y_2 = \varepsilon^2 y \tag{41}$$

This gives for the derivatives

$$\frac{\partial}{\partial t} \rightarrow \frac{\partial}{\partial t} + \varepsilon \frac{\partial}{\partial T_1} + \varepsilon^2 \frac{\partial}{\partial T_2} \tag{42}$$

$$\frac{\partial}{\partial y} \rightarrow \frac{\partial}{\partial y} + \varepsilon \frac{\partial}{\partial Y_1} + \varepsilon^2 \frac{\partial}{\partial Y_2} \tag{43}$$

The solution is adopted to be of the form

$$\phi = \phi^{(1)} + \varepsilon \phi^{(2)} + \varepsilon^2 \phi^{(3)} + \cdots \tag{44}$$

We denote the linear part of the operator in the equation by

$$L \equiv \left(\frac{\partial}{\partial t} + U(x) \frac{\partial}{\partial y} \right) \nabla^2 + \left(\beta - \frac{d^2 U(x)}{dx^2} \right) \frac{\partial}{\partial y} \tag{45}$$

Substituting the Eq.(44) and the forms of the derivation operators Eqs.(42, 43) in the equation of motion we obtain from the equality between the coefficients of the powers of the variable ε:

$$L\phi^{(1)} = 0 \tag{46}$$

$$\begin{aligned} L\phi^{(2)} &= \\ &= \left(\frac{\partial}{\partial T_1} + U \frac{\partial}{\partial Y_1} \right) \nabla^2 \phi^{(1)} + (\beta - U'') \frac{\partial \phi^{(1)}}{\partial Y_1} \\ &\quad -2 \left(\frac{\partial}{\partial t} + U \frac{\partial}{\partial y} \right) \frac{\partial^2 \phi^{(1)}}{\partial y \partial Y_1} \\ &\quad - \left[\phi^{(1)}, \nabla^2 \phi^{(1)} \right] \end{aligned} \tag{47}$$

$$\begin{aligned} L\phi^{(3)} &= \\ &= \left(\frac{\partial}{\partial T_2} + U \frac{\partial}{\partial Y_2} \right) \nabla^2 \phi^{(1)} - (\beta - U'') \frac{\partial \phi^{(1)}}{\partial Y_2} \\ &\quad -2 \left(\frac{\partial}{\partial t} + U \frac{\partial}{\partial y} \right) \frac{\partial^2 \phi^{(1)}}{\partial y \partial Y_2} \\ &\quad - \left(\frac{\partial}{\partial t} + U \frac{\partial}{\partial y} \right) \frac{\partial^2 \phi^{(1)}}{\partial Y_1^2} \\ &\quad -2 \left(\frac{\partial}{\partial T_1} + U \frac{\partial}{\partial Y_1} \right) \frac{\partial^2 \phi^{(1)}}{\partial y \partial Y_1} \\ &\quad - \left(\frac{\partial}{\partial T_1} + U \frac{\partial}{\partial Y_1} \right) \nabla^2 \phi^{(2)} \\ &\quad -2 \left(\frac{\partial}{\partial t} + U \frac{\partial}{\partial y} \right) \frac{\partial^2 \phi^{(2)}}{\partial y \partial Y_1} - (\beta - U'') \frac{\partial \phi^{(2)}}{\partial Y_1} \\ &\quad - \frac{\partial \phi^{(2)}}{\partial Y_1} \frac{\partial}{\partial x} \nabla^2 \phi^{(1)} + \frac{\partial \phi^{(1)}}{\partial x} \frac{\partial}{\partial Y_1} \nabla^2 \phi^{(1)} \\ &\quad -2 \left(\frac{\partial \phi^{(1)}}{\partial y} \frac{\partial^3 \phi^{(1)}}{\partial y \partial x \partial Y_1} - \frac{\partial \phi^{(1)}}{\partial x} \frac{\partial^3 \phi^{(1)}}{\partial y^2 \partial Y_1} \right) \\ &\quad - \left[\phi^{(1)}, \nabla^2 \phi^{(2)} \right] - \left[\phi^{(2)}, \nabla^2 \phi^{(1)} \right] \end{aligned} \tag{48}$$

The solution is represented on the space and time scales.

3.1 The Enveloppe Equation

The solution of the linear part of the operator, Eq.(46) is adopted as the superposition of two propagating waves

$$\begin{aligned}\phi^{(1)} &= A_1(T_1,T_2,Y_1,Y_2)\,\varphi_1(x)\exp(ik_1y-i\omega_1t) \\ &\quad +A_2(T_1,T_2,Y_1,Y_2)\,\varphi_2(x)\exp(ik_2y-i\omega_2t) \\ &\quad +c.c.\end{aligned} \tag{49}$$

The two functions φ_n, $n=1,2$, are solutions of the equation

$$\frac{d^2\varphi_n}{dx^2}-\left(k_n^2-\frac{\beta-U''}{U-c_n}\right)\varphi_n=0 \tag{50}$$

with

$$\varphi_n(0)=\varphi_n(L)=0$$

In this formula, $(0,L)$ are the limits of the spatial domain in the minor radius direction where the flow $U(x)\neq 0$; $c_n=\omega_n/k_n$ are the phase velocities. To find the solutions of this equation we have to use numerical methods. This is a Schrodinger-type equation where the potential depends on the energy, (since the phase velocity c_n depends on the wavenumber k_n). We must use numerical methods suitable for Sturm-Liouville equations and in general the solution exists for only particular values of the parameters. We choose to fix the frequencies ω_n and consider k_n as the parameter to be determined. The problem is complicated by the possible occurence of singularities in the potential due to resonances between the phase velocity and the fixed zonal flow. We will avoid any resonance since the physical content of the processes consisting of the direct energy transfer between the fluctuations and the flow represents a different physical problem and requires separate consideration. In the present work we examine the effect of the turbulence on the poloidal rotation as being due to the nonuniform diffusion rates and Stringer mechanism.

The parameters in the Eqs.(50) are β, ε and the sheared flow velocity $U(x)$. We take for $U(x)$ a symmetric Gaussian-like profile shifted in amplitude and retain its maximum, half-width and asymptotic value (*i.e.* at $x=0$ or L) as parameters. The two frequencies have fixed values during the eigenvalue search but they must be considered free parameters as well. This builds up a large space of parameters which should be sampled. For some point in this space the eigenvalue problems Eqs.(50) are solved and the phase velocities compared with the profile of $U(x)$. The presence of resonances renders the set of parameters useless. Since (as will result from the formuals below) there are also other possible resonances, the determination of a useful solution (k_1,k_2) is difficult and requires many trials.

The numerical methods that have been used in solving Eqs.(50) belong to two classes: boundary value integration methods and respectively shooting methods. When the sign of the "potential" in the equation is negative everywhere on $(0,L)$, which occurs for short wavelengths and smooth fluid velocity profile, the boundary conditions cannot be fulfilled except for asymtotically, *i.e.* assuming arbitrary small nonzero values at these limits, since the solution decays exponentially. However most of the boundary value integrators we have tried gave amplitudes of magnitudes comparable to the boundary value, which essentially are homotopic to the trivial vanishing solution (the equation is homogeneous) and render useless this approach. We have to use a shooting method accepting the difficulty that the initial value of the parameter to be determined (eigenvalue) must be placed not far from the final value. We have used the specialized package SLEIGN2 interactively. For the investigation of ranges of parameters most of the calculations have been performed using the routine D02KEF of NAG. For identical conditions the two numerical codes gave the same results within the accepted tolerance.

Using the routine **D02KEF** we divide the integration range $(0, L)$ into four intervals corresponding roughly to a different behaviour of the potential. This should simplify the task of the integrator. A single matching point is assumed, usually at the centre. The tolerence is 10^{-8} which however does not exclude fake convergence in case of an initialisation of the eigenvalue very far from the true solution.

We notice from Eqs.(50) that the potential can easily be dominated by high shear U'' especially for smaller wavenumbers k_n. Actually we are more interested to investigate regimes where only mild effects of the seed sheared rotation arise during the process of self-modulation since, as explained before, the possible evolution to the nonuniformity in the poloidal direction will be the source of the torque on the plasma. Avoiding regimes of U''-dominated potentials we assume low shear and take the Gaussian profile with $U_{\max} - U_{\min} = 2 \times 10^4$ (m/s) represented in Figure (1). The following set of physical parameters has been used: minor radius $a = 1$ m, magnetic field $B_T = 3.5$ T, electron temperature $T_e = 1$ KeV, ion temperature $T_i = 1$ KeV, density $n = 10^{20}$ m^{-3}, $L_n = 10$ m. The parameters in the barotropic equation are then: $\beta = 33.526$ and $\varepsilon = 0.28571$. The two frequencies are $\omega_1 = 85 \times 10^2$ s^{-1} and $\omega_2 = 85 \times 10^4$ s^{-1}. It is assumed that the space region of significant magnitude of $U(x)$ is $L = 0.1$ m. The following two eigenvalues are obtained, normalised to L^{-1}: $k_1 = 0.5525$ and $k_2 = 3.2029$. The potentials in the Schrodinger-like equations and the eigenfunctions are shown in Figure (2).

The two eigenfunctions $\varphi_{1,2}$ have an important role in the following steps of the calculation. Substituting the Eq.(49) in the barotropic equation and expanding the operators, we obtain:

$$\begin{aligned} L\phi^{(2)} = & -\sum_{n=1}^{2} G_{1n} \exp\left(ik_n y - i\omega_n t\right) \\ & -\sum_{n=1}^{2} ig_{2n} A_n^2 \exp\left[i2\left(k_n y - \omega_n t\right)\right] \\ & -g_3 A_1 A_2 \exp\left[i\left(k_{12} y - \omega_{12} t\right)\right] \\ & -ig_4 A_1 A_2^* \exp\left[i\left(\alpha_{12} y - \sigma_{12} t\right)\right] \\ & +c.c. \end{aligned} \tag{51}$$

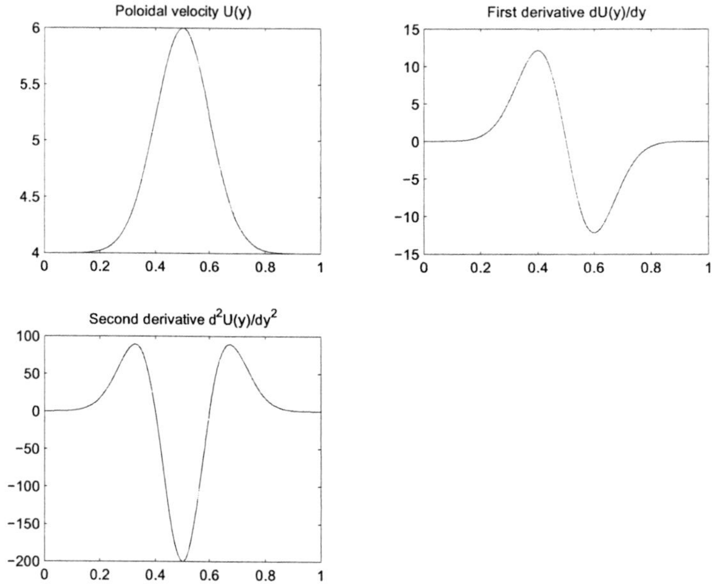

Figure 1: Profile of the seed sheared poloidal velocity. Also are plotted the first and the second derivatives.

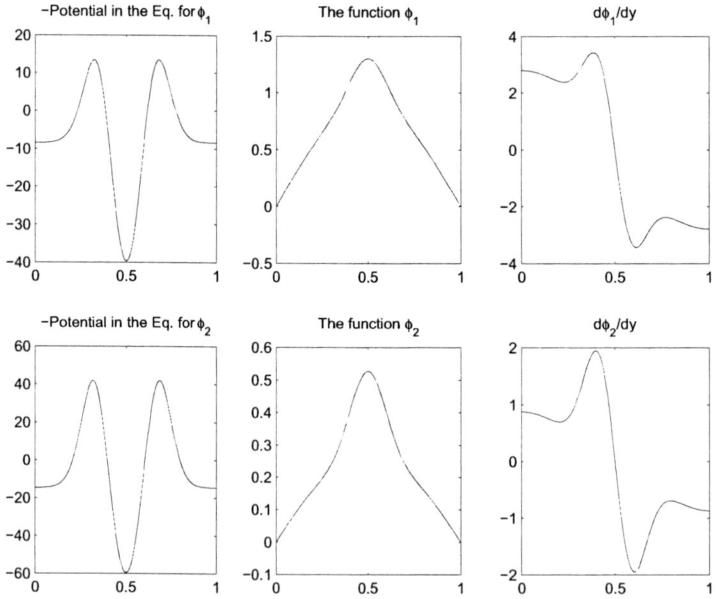

Figure 2: The potentials in the Schrodinger-type equations and the eigenfunctions $\varphi_{1,2}$. The derivatives of the eigenfunctions are aslo plotted.

where

$$\begin{aligned} k_{12} &= k_1 + k_2 \\ \omega_{12} &= \omega_1 + \omega_2 \end{aligned} \tag{52}$$

We have $k_{12} = 3.755$, $\omega = 8.585$.

$$\begin{aligned} \alpha_{12} &= k_1 - k_2 \\ \sigma_{12} &= \omega_1 - \omega_2 \end{aligned} \tag{53}$$

or: $\alpha_{12} = -2.650$, $\sigma_{12} = -8.415$.

$$\begin{aligned} G_{1n} &= \left(\frac{d^2 \varphi_n}{dx^2} - k_n^2 \varphi_n \right) \frac{\partial A_n}{\partial T_1} + \\ &+ \left[U \left(\frac{d^2 \varphi_n}{dx^2} - k_n^2 \varphi_n \right) + \left(\beta - U'' \right) \varphi_n - 2k_n \left(U - c_n \right) \right] \frac{\partial A_n}{\partial Y_1} \end{aligned} \tag{54}$$

$$g_{2n} = k_n \left(\varphi_n \frac{d}{dx} - \frac{d\varphi_n}{dx} \right) \frac{d^2 \varphi_n}{dx^2} \tag{55}$$

$$\begin{aligned} g_3 &= \left(k_1 \varphi_1 \frac{d}{dx} - k_2 \frac{d\varphi_1}{dx} \right) \left(\frac{d^2 \varphi_2}{dx^2} - k_2^2 \varphi_2 \right) \\ &+ \left(k_2 \varphi_2 \frac{d}{dx} - k_1 \frac{d\varphi_2}{dx} \right) \left(\frac{d^2 \varphi_1}{dx^2} - k_1^2 \varphi_1 \right) \end{aligned} \tag{56}$$

$$\begin{aligned} g_4 &= \left(k_1 \varphi_1 \frac{d}{dx} + k_2 \frac{d\varphi_1}{dx} \right) \left(\frac{d^2 \varphi_2}{dx^2} - k_2^2 \varphi_2 \right) \\ &+ \left(k_2 \varphi_2 \frac{d}{dx} + k_1 \frac{d\varphi_2}{dx} \right) \left(\frac{d^2 \varphi_1}{dx^2} - k_1^2 \varphi_1 \right) \end{aligned} \tag{57}$$

These functions can be obtained after calculating the eigenfunctions $\varphi_{1,2}$ and are represented in Figure (3).

Various solutions to the equation (51) are possible, taking the space-time dependence similar to one of the terms which compose the right hand side.

The second term suggests solutions of the type

$$\phi_{2n}^{(2)} = W_{2n} A_n^2 \exp \left[i2 \left(k_n y - \omega_n t \right) \right] + c.c. \tag{58}$$

The third term suggets solutions of the type

$$\phi_3^{(2)} = W_3 A_1 A_2 \exp \left[i \left(k_{12} y - \omega_{12} t \right) \right] + c.c. \tag{59}$$

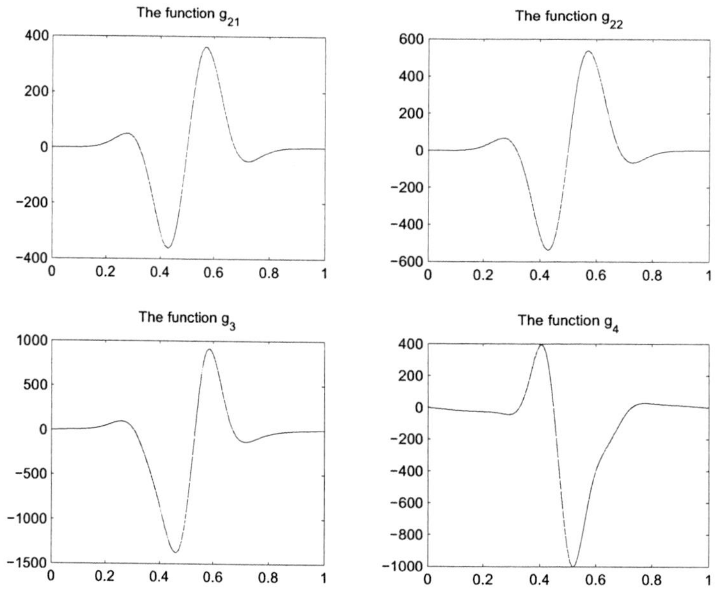

Figure 3: Graphs of the functions g's.

The fourth term gives

$$\phi_4^{(2)} = W_4 A_1 A_2^* \exp\left[i\left(\alpha_{12} y - \sigma_{12} t\right)\right] + c.c. \tag{60}$$

In the formulas above the functions $W_{2n}(x)$ are solutions of the equations:

$$\begin{aligned} \frac{d^2 W_{2n}}{dx^2} - \left(4k_n^2 - \frac{\beta - U''}{U - c_n}\right) W_{2n} &= \frac{g_{2n}}{2\left(Uk_n - \omega_n\right)} \\ \text{with } W_{2n}(0) &= W_{2n}(L) = 0 \end{aligned} \tag{61}$$

$$\begin{aligned} \frac{d^2 W_3}{dx^2} - \left(k_{12}^2 - \frac{k_{12}\left(\beta - U''\right)}{Uk_{12} - \omega_{12}}\right) W_3 &= \frac{g_3}{Uk_{12} - \omega_{12}} \\ \text{with } W_3(0) &= W_3(L) = 0 \end{aligned} \tag{62}$$

$$\begin{aligned} \frac{d^2 W_4}{dx^2} - \left(\alpha_{12}^2 - \frac{\alpha_{12}\left(\beta - U''\right)}{U\alpha_{12} - \sigma_{12}}\right) W_4 &= \frac{g_4}{U\alpha_{12} - \sigma_{12}} \\ \text{with } W_4(0) &= W_4(L) = 0 \end{aligned} \tag{63}$$

These equations are integrated numerically with a boundary value integrator, a finite difference method. The results are shown in Figures (4), (5) and (6). In the following figures are shown the functions W_{1n} which will be introduced in Eq.(70).

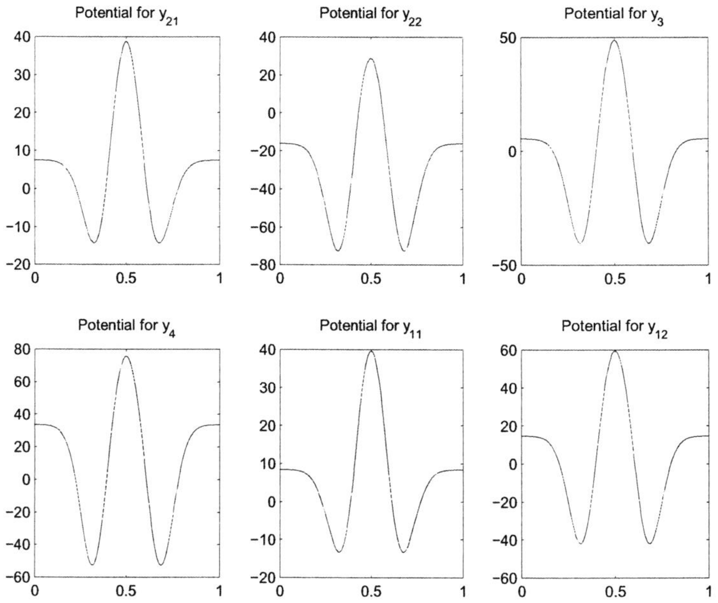

Figure 4: Graphs of the potentials occuring in the equations for the functions w's.

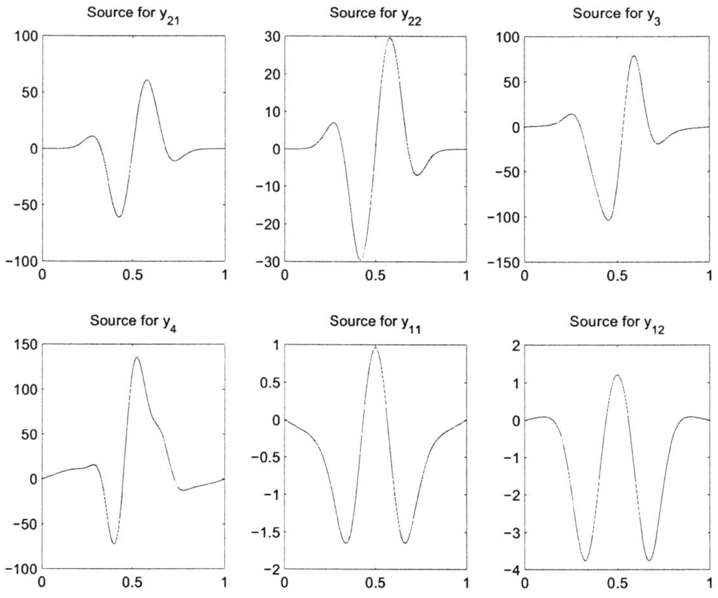

Figure 5: Graphs of the source terms in the equations for the functions w's.

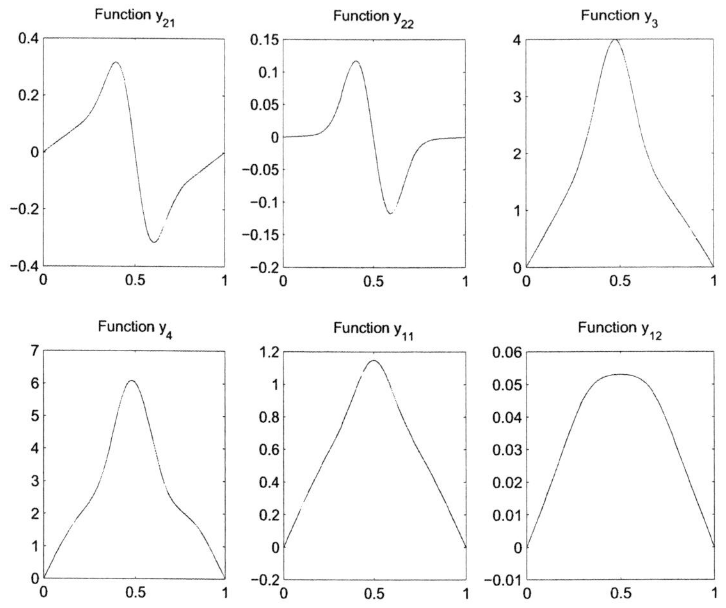

Figure 6: Graphs of the functions w's.

The first term gives a solution of the following form

$$\phi_{1n}^{(2)} = \varphi_{1n}^{(2)} \exp\left(ik_n y - i\omega_n t\right) + c.c. \tag{64}$$

where the function $\varphi_{1n}^{(2)}$ satisfies the equation

$$\begin{aligned} \frac{d^2\varphi_{1n}^{(2)}}{dx^2} - \left(k_n^2 - \frac{\beta - U''}{U - c_n}\right)\varphi_{1n}^{(2)} &= \\ &= \frac{i}{k_n\left(U - c_n\right)}\left\{\left(\frac{d^2\varphi_n}{dx^2} - k_n^2\varphi_n\right)\frac{\partial A_n}{\partial T_1}\right. \\ &\quad + \left[U\left(\frac{d^2\varphi_n}{dx^2} - k_n^2\varphi_n\right) + \left(\beta - U''\right)\varphi_n\right. \\ &\quad \left.\left. -2k_n^2\left(U - c_n\right)\right]\frac{\partial A_n}{\partial Y_1}\right\} \end{aligned} \tag{65}$$

This equation serves to generate a condition on the two amplitudes $A_{1,2}$.

When the left hand side of the equation is multiplied by the function φ_n and integrated between the limits on the x domain, it gives zero, due to the boundary conditions. The same must then be true for the right hand side. This gives a "solubility condition", on the space time scales slower at order 1:

$$\frac{\partial A_n}{\partial T_1} + c_{gn}\frac{\partial A_n}{\partial Y_1} = 0 \tag{66}$$

where

$$c_{gn} = c_n + 2\int_0^L k_n^2 \varphi_n^2 dx \left[\int_0^L \frac{\beta - U''}{(U - c_n)^2}\varphi_n^2 dx\right]^{-1} \tag{67}$$

Expressed in units of length L and units of time $t_0 = 10^{-4}$, the *group velocities* are : $c_{g1} = 0.30328$ and $c_{g2} = 3.9908$.

Then it is clear that the amplitude A_n propagates at the speed c_{gn} :

$$A_n = A_n\left(Y_1 - c_{gn}T_1, Y_2, T_2\right) \tag{68}$$

Now we introduce a new function

$$\varphi_{1n}^{(2)} = iW_{1n}\frac{\partial A_n}{\partial Y_1} \tag{69}$$

The equation satisfied by W_{1n} is

$$\begin{aligned} &\frac{d^2 W_{1n}}{dx^2} - \left(k_n^2 - \frac{\beta - U''}{u - c_n}\right) W_{1n} \\ &= \frac{1}{k_n\left(U - c_n\right)}\left[\left(c_n - c_{gn}\right)\left(\frac{d^2\varphi_n}{dx^2} - k_n^2\varphi_n\right) - 2k_n^2\left(U - c_n\right)\varphi_n\right] \end{aligned} \tag{70}$$

with the bounday conditions

$$W_{1n}(0) = W_{1n}(L) = 0 \tag{71}$$

At this order (2) there exist the following solutions

$$\begin{aligned} \phi^{(2)} \;\; = \;\; & \phi_{11}^{(2)} + \phi_{12}^{(2)} + \phi_{21}^{(2)} + \phi_{22}^{(2)} \\ & + \phi_3^{(2)} + \phi_4^{(2)} + \\ & + \Phi\left(x, T_1, Y_1\right) \end{aligned} \tag{72}$$

The last function will be determined by the condition of solubility at the next order on the space-time scales $(O(\varepsilon^2))$. To obtain the evolution equations for Φ and A_n, we use the expressions for $\phi^{(1)}$ Eq.(49) and $\phi^{(2)}$, Eq.(72) in the right hand side of the equation (48), and all inhomogeneous terms are obtained. In additions there are terms which are not dependent on y and t, and **these terms should be zero**. This leads to the solubility condition relating the function Φ (correction to the mean flow due to the presence of the wave packets) to the amplitude A_n.

$$\begin{aligned} &\left(\frac{\partial}{\partial T_1} + U\frac{\partial}{\partial Y_1}\right)\frac{\partial^2\Phi}{\partial x^2} + \left(\beta - U''\right)\frac{\partial\Phi}{\partial Y_1} \\ &= \sum_{n=1}^{2}\left\{\frac{d^2}{dx^2}\left[k_n\left(W_{1n}\frac{d\varphi_n}{dx} - \varphi_n\frac{dW_{1n}}{dx}\right)\right] - 4k_n^2\varphi_n\frac{d\varphi_n}{dx}\right. \\ &\left. + \left(\varphi_n\frac{d}{dx} - \frac{d\varphi_n}{dx}\right)\left(\frac{d^2\varphi_n}{dx^2} - k_n^2\varphi_n\right)\right\}\frac{\partial}{\partial Y_1}\left|A_n\right|^2 \end{aligned} \tag{73}$$

This relation will be used later.

Returning to the equation (48), we examine the content of the inhomogeneous part (the right hand side). There are terms which have a space-time dependence

$$\exp\left[ik_n y - i\omega_n t\right] \tag{74}$$

and in order to have solutions of this type, we need to impose solubility conditions, otherwise there will arise secular terms at the order $\phi^{(3)}$. To find these restrictions (solubility conditions) we multiply the equation with the function conjugated to (74)

$$\exp\left[-ik_n y + i\omega_n t\right]$$

and integrate time on the interval $(0, 2\pi/\omega_n)$ and space: y on the interval $(0, 2\pi/k_n)$ and x on the interval $(0, L)$. On the left hand side we obtain zero, so this is what we must obtain also from the right hand side. The conditions which results are:

$$\left(\frac{\partial}{\partial T_2} + c_{g1}\frac{\partial}{\partial Y_2}\right) A_1 - i\alpha_1 \frac{\partial^2 A_1}{\partial Y_1^2} - i\left(\rho_1 \left|A_1\right|^2 + \gamma_{12}\left|A_2\right|^2 + \lambda_1\right) A_1 = 0 \tag{75}$$

$$\left(\frac{\partial}{\partial T_2} + c_{g2}\frac{\partial}{\partial Y_2}\right) A_2 - i\alpha_2 \frac{\partial^2 A_2}{\partial Y_1^2} - i\left(\rho_2 \left|A_2\right|^2 + \gamma_{21}\left|A_1\right|^2 + \lambda_2\right) A_2 = 0 \tag{76}$$

The notations which have been introduced are: $\alpha_n = \frac{I_{n0}}{\Pi_n}$, $\rho_n = \frac{I_{nn}}{\Pi_n}$, $\gamma_{12} = \frac{I_{12}}{\Pi_1}$, $\gamma_{21} = \frac{I_{21}}{\Pi_2}$, $\lambda_n = \frac{I_{n3}}{\Pi_n}$ and

$$\Pi_n = -\int_0^L \frac{\varphi_n}{U - c_n} f_n dx\ ,\ I_{n0} = \int_0^L \frac{\varphi_n}{U - c_n} f_{n0} dx \tag{77}$$

$$I_{nn} = \int_0^L \frac{\varphi_n}{U - c_n} f_{nn} dx\ ,\ I_{12} = \int_0^L \frac{\varphi_1}{U - c_1} f_{12} dx \tag{78}$$

$$I_{21} = \int_0^L \frac{\varphi_2}{U - c_2} f_{21} dx\ ,\ I_{n3} = \int_0^L \frac{\varphi_n}{U - c_n} f_{n3} dx \tag{79}$$

The equations satisfied by all these functions are given given in Ref.[9].

$$f_n = -\left(\frac{d^2\varphi_n}{dx^2} - k_n^2 \varphi_n\right) \tag{80}$$

$$h_n = -U\left(\frac{d^2\varphi_n}{dx^2} - k_n^2\varphi_n\right) - \left(\beta - U''\right)\varphi_n + 2k_n^2\left(U - c_n\right) \tag{81}$$

$$\begin{aligned} f_{n0} &= -2k_n\left(U - c_{gn}\right)\varphi_n - k_n\left(U - c_n\right) \\ &\quad -\left(U - c_{gn}\right)\left(\frac{d^2 W_{1n}}{dy^2} - k_n^2 W_{1n}\right) \\ &\quad +2k_n^2\left(U - c_n\right)W_{1n} - \left(\beta - U''\right)W_{1n} \end{aligned} \tag{82}$$

$$\begin{aligned} f_{nn} &= -\left(2k_n W_{2n}\frac{d}{dx} + k_n\frac{dW_{2n}}{dx}\right)\left(\frac{d^2\varphi_n}{dx^2} - k_n^2\varphi_n\right) \\ &+ \left(k_n\varphi_n\frac{d}{dx} + 2k_n\frac{d\varphi_n}{dx}\right)\left(\frac{d^2W_{2n}}{dx^2} - 4k_n^2 W_{2n}\right) \end{aligned} \tag{83}$$

$$\begin{aligned} f_{12} &= -\left(k_{12}W_3\frac{d}{dx} + k_2\frac{dW_3}{dx}\right)\left(\frac{d^2\varphi_2}{dx^2} - k_2^2\varphi_2\right) \\ &+ \left(k_2\varphi_2\frac{d}{dx} + k_{12}\frac{d\varphi_2}{dx}\right)\left(\frac{d^2W_3}{dx^2} - k_{12}^2 W_3\right) \\ &+ \left(k_2\varphi_2\frac{d}{dx} - \alpha_{12}\frac{d\varphi_2}{dx}\right)\left(\frac{d^2W_4}{dx^2} - \alpha_{12}^2 W_4\right) \\ &- \left(\alpha_{12}W_4\frac{d}{dx} - k_2\frac{dW_4}{dx}\right)\left(\frac{d^2\varphi_2}{dx^2} - k_2^2\varphi_2\right) \end{aligned} \tag{84}$$

$$\begin{aligned} f_{21} &= -\left(k_{12}W_3\frac{d}{dx} + k_1\frac{dW_3}{dx}\right)\left(\frac{d^2\varphi_1}{dx^2} - k_1^2\varphi_1\right) \\ &- \left(k_1\varphi_1\frac{d}{dx} + \alpha_{12}\frac{d\varphi_1}{dx}\right)\left(\frac{d^2W_4}{dx^2} - \alpha_{12}^2 W_4\right) \\ &+ \left(k_1\varphi_1\frac{d}{dx} + k_{12}\frac{d\varphi_2}{dx}\right)\left(\frac{d^2W_3}{dx^2} - k_{12}^2 W_3\right) \\ &+ \left(\alpha_{12}W_4\frac{d}{dx} + k_1\frac{dW_4}{dx}\right)\left(\frac{d^2\varphi_1}{dx^2} - k_1^2\varphi_1\right) \end{aligned} \tag{85}$$

$$f_{n3} = -k_n\left(\varphi_n\frac{\partial^3\Phi}{\partial x^3} - \left(\frac{d^2\varphi_n}{dx^2} - k_n^2\varphi_n\right)\frac{\partial\Phi}{\partial x}\right) \tag{86}$$

We use a boundary value integrator for the inhomogeneous equations and (the NAG routine **D02JAF**). The results are given in the Figures (7) and (8).

After numerical integrations the follwing values are obtained for the constants: $\alpha_1 = 0.39976$, $\alpha_2 = 0.45246$, $\rho_1 = -0.10182$, $\rho_2 = 1.8115$, $\gamma_{12} = 0.091474$, $\gamma_{21} = -5.3768$.

The following transformation (Jeffrey and Kawahara) is performed in Ref.[9] :

$$\begin{aligned} T &= T_2 \\ X &= \frac{1}{\varepsilon}(Y_2 - c_{g1}T_2) = Y_1 - c_{g1}T_1 \end{aligned} \tag{87}$$

Then the equation connecting Φ and A_n becomes

$$\begin{aligned} &\frac{\partial}{\partial Y}\left[(U - c_{g1})\frac{\partial^2\Phi}{\partial x^2} + (\beta - U'')\,\Phi\right] \\ &= \sum_{n=1}^{2}\left[k_n\frac{d^2}{dx^2}\left(W_{1n}\frac{d\varphi_n}{dx} - \varphi_n\frac{dW_{1n}}{dx}\right) - 4k_n\varphi_n\frac{d\varphi_n}{dx}\right. \\ &\left. - \left(\varphi_n\frac{d}{dx} - \frac{d\varphi_n}{dx}\right)\left(\frac{d^2\varphi_n}{dx^2} - k_n^2\varphi_n\right)\right]\frac{\partial}{\partial Y}|A_n|^2 \end{aligned} \tag{88}$$

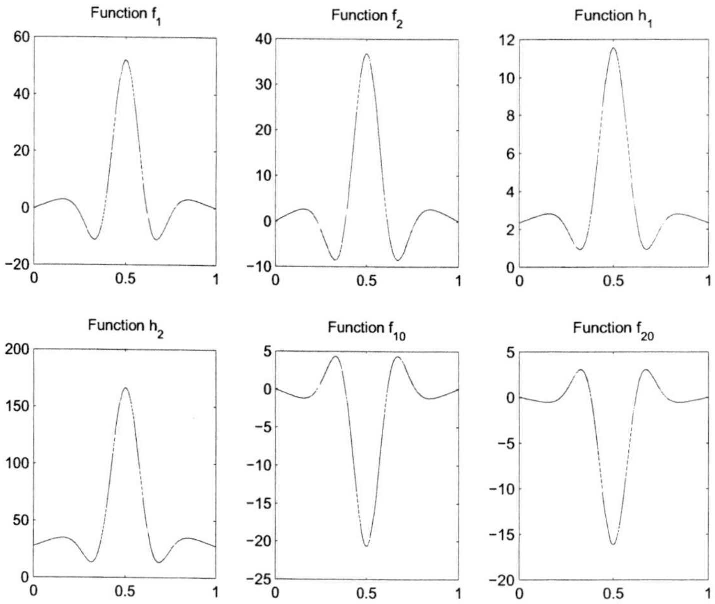

Figure 7: Graphs of the first group of functions f.

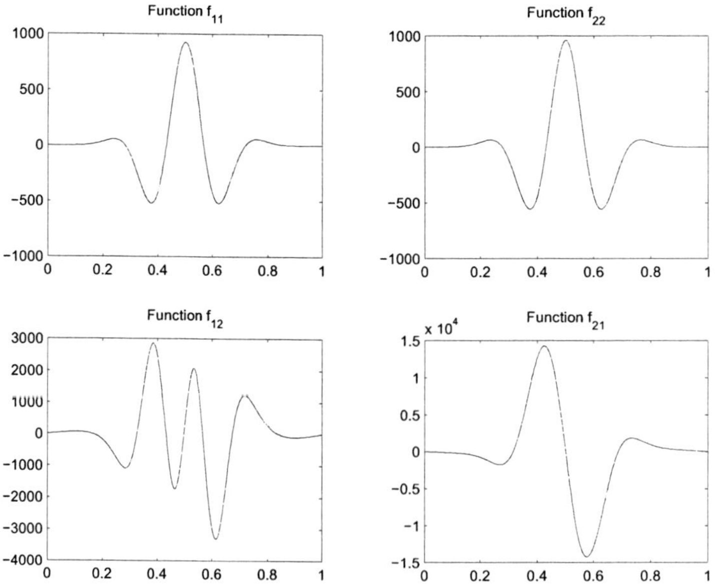

Figure 8: Graphs of the second group of functions f.

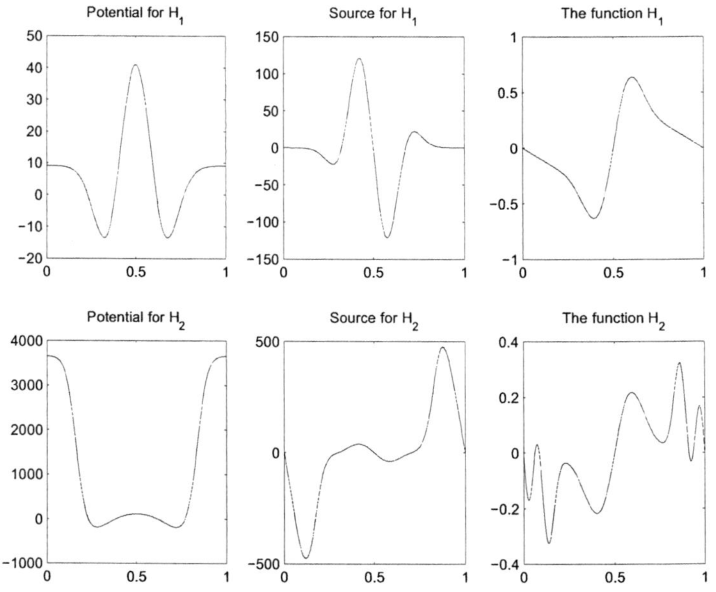

Figure 9: Graphs of the functions H.

The solution of the equation (88) is considered to be of the form

$$\Phi = \sum_{n=1}^{2} H_n(x) |A_n|^2 \tag{89}$$

and the equation for $H_n(x)$ results:

$$\begin{aligned} & (U - c_{g1}) \frac{\partial^2 H_n}{\partial x^2} + (\beta - U'') H_n \\ = \; & k_n \frac{d^2}{dx^2} \left(W_{1n} \frac{d\varphi_n}{dx} - \varphi_n \frac{dW_{1n}}{dx} \right) - 4 k_n \varphi_n \frac{d\varphi_n}{dx} \\ & - \left(\varphi_n \frac{d}{dx} - \frac{d\varphi_n}{dx} \right) \left(\frac{d^2 \varphi_n}{dx^2} - k_n^2 \varphi_n \right) \end{aligned} \tag{90}$$

The change of variables is performed in the equations for the amplitudes A_n and, again, the Eq.(89) is used:

$$i \frac{\partial A_1}{\partial T} + \alpha_1 \frac{\partial^2 A_1}{\partial Y^2} + \left(\sigma_1 |A_1|^2 + \nu_{12} |A_2|^2 \right) A_1 = 0 \tag{91}$$

$$i \left(\frac{\partial}{\partial T} - \delta \frac{\partial}{\partial Y} \right) A_2 + \alpha_2 \frac{\partial^2 A_2}{\partial Y^2} + \left(\sigma_2 |A_2|^2 + \nu_{21} |A_1|^2 \right) A_2 = 0 \tag{92}$$

where

$$\delta = \frac{1}{\varepsilon} (c_{g1} - c_{g2}) \tag{93}$$

$$\sigma_n = \rho_n - \frac{1}{\Pi_n}\int_0^L \frac{k_n\varphi_n}{U-c_n}\left[\varphi_n\frac{d^3H_n}{dx^3} - \left(\frac{d^2\varphi_n}{dx^2} - k_n^2\varphi_n\right)\frac{dH_n}{dx}\right]dx \tag{94}$$

$$\nu_{12} = \gamma_{12} - \frac{1}{\Pi_1}\int_0^L \frac{k_1\varphi_1}{U-c_1}\left[\varphi_1\frac{d^3H_2}{dx^3} - \left(\frac{d^2\varphi_1}{dx^2} - k_1^2\varphi_1\right)\frac{dH_2}{dx}\right]dx \tag{95}$$

$$\nu_{21} = \gamma_{21} - \frac{1}{\Pi_2}\int_0^L \frac{k_2\varphi_2}{U-c_2}\left[\varphi_2\frac{d^3H_1}{dx^3} - \left(\frac{d^2\varphi_2}{dx^2} - k_2^2\varphi_2\right)\frac{dH_1}{dx}\right]dx \tag{96}$$

The following transformation is convenient:

$$A_2 = B\exp\left[i\frac{\delta}{2\alpha_2}\left(Y + \frac{\delta}{2}T\right)\right] \tag{97}$$

This leads to the following equations:

$$i\frac{\partial A_1}{\partial T} + \alpha_1\frac{\partial^2 A_1}{\partial Y^2} + \left(\sigma_1\left|A_1\right|^2 + \nu_{12}\left|B\right|^2\right)A_1 = 0 \tag{98}$$

$$i\frac{\partial B}{\partial T} + \alpha_2\frac{\partial^2 B}{\partial Y^2} + \left(\sigma_2\left|B\right|^2 + \nu_{21}\left|A_1\right|^2\right)B = 0 \tag{99}$$

The constants are: $\sigma_1 = -1.6324$, $\sigma_2 = 1.9331$, $\nu_{12} = -0.53387$, $\nu_{21} = -2.9014$. The constants $\alpha_{1,2}$ are called the "dispersion", $\sigma_{1,2}$ are called the nonlinearity ("Landau") constants ; $\nu_{12,21}$ represent the coupling of the two amplitudes.

Numerical experience shows that the dispersion constants are sensitive to the precision of the integration scheme. We use for **D02KEF** the value for tolerence 10^{-8}.

The system of coupled Cubic Nonlinear Schrodinger Equations has been integrated numerically in relation with many applications [11], [12]. One important conclusion is that, in the case of a single equation, the sign of the product of the dispersion and of the nonlinearity parameters establishes the possibility of a soliton formation.

3.2 Numerical Search for Classes of Admissible Envelope Solutions

In principle an interactive search for eigenvalues (either with SLEIGN or **D02KEF**) is able to obtain the eigenvalues k_1, k_2 eventually after series of less-inspired initializations. However in many situations resonances occur and the sources and/or the potentials in the differential equations for the functions W_n, f_n become discontinuous. Since these eigenvalues cannot be used we have to devise an automatic search over the space of parameters with checks for continuity of the functions and automatic discarding of wrong results. This is accomplished by the routine **P01ABF** of NAG.

The search reveals that the short and respectively long *radial* wavelength part of the ion wave spectrum have different behavior under the modulation instability. The most affected is the long wavelength part (small k_x) with ω of the order of the ion-diamagnetic frequency. This corresponds to the dominant part of the ion spectrum, because, as can be seen in numerous numerical simulations, the dominant structures are elongated in the radial

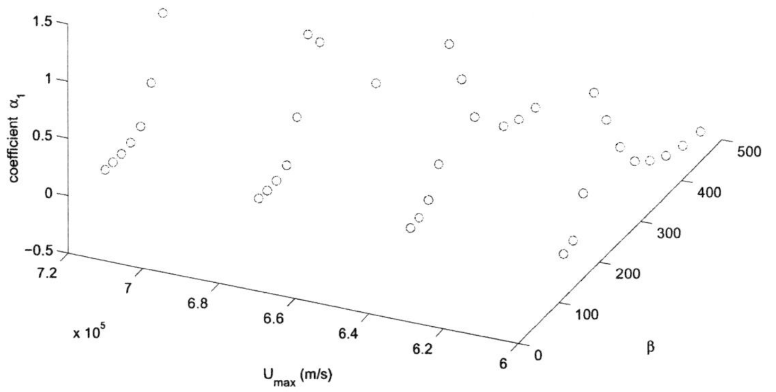

Figure 10: Parameter dependence of the coefficient α_1.

direction, with various degrees of tilting compared to the radius. It is also important to note that finding the eigenvalues in the scan of the parameter space was possible in the condition that the eigenfunctions $\varphi_{1,2}(x)$ have low number of oscillations in the radial direction (one or two). Again this corresponds to the important part of the ITG turbulence spectrum.

The variation of the parameters of the system of NSE with the physical parameters of the problem can be studied by the automatic search of eigenvalues, as described before. Below are plotted four graphs showing this dependence.

4 Exact Solutions of the Nonlinear Schrodinger Equation and the Nonlinear Stability Problem

4.1 Introduction

The Nonlinear Schrodinger Equation can be solved *exactly* on an infinite spatial domain using the Inverse Scattering Transform. The first step is to represent the nonlinear equation as a condition of compatibility of two *linear* equations. This is achieved by finding a pair of linear operators (Lax operators) and noticing that for the first of them the eigenfunction problem is a Schrodinger equation leading to a quantum scattering problem. The unknown function of the NSE is the potential of the Schrodinger equation. Using the initial condition of the NSE the scattering data are determined at $t = 0$. The scattering data have simple time dependence and so they can be evloved in time to the desired moment. Returning from this new set of scattering data (to the potential that has produced it) is achieved by solving an integral equation (Gelfand-Levitan-Marchenko) and provides the solution of the NSE. In general this solution consists of solitons and "radiation".

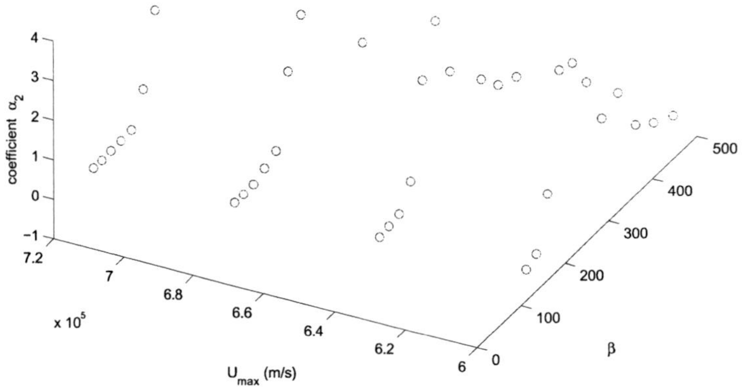

Figure 11: Parameter dependence of the coefficient α_2.

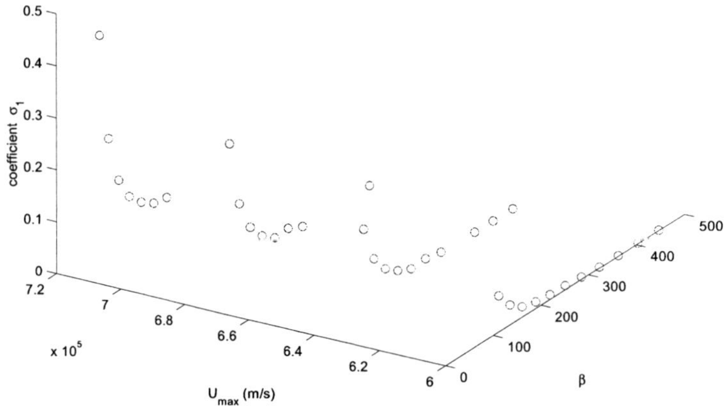

Figure 12: Parameter dependence of the coefficient σ_1.

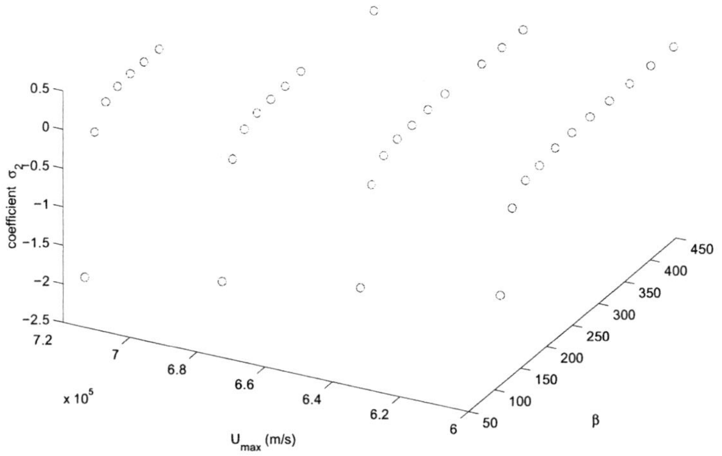

Figure 13: Parameter dependence of the coefficient σ_2.

The Inverse Scattering Transform (IST) on periodic domain [15], [14], [17] is a more powerful procedure since it reveals and takes advantage of the deep geometric and topological nature of the problem. The admissible solutions of the Lax eigenvalue problem can now only be periodic functions. Since after one period the change of any solution can only be linear (via the *monodromy* matrix) the periodic or antiperiodic functions must be found as eigenfunctions of this monodromy operator (matrix), under the constrain that the eigenvalues are complex of unit absolute value. This singles out a set of complex values of the spectral parameter λ (the formal eigenvalue in the Lax equation), the *main spectrum*. From this set, the particular values that make the two eigenfunctions to coincide are called *non-degenerate*. Here the squared Wronskian (which is a space and time invariant) has zeros on the complex λ plane. Then the non-degenerate λ's are singular points for the complex-valued Wronskian. Removing the square-root indeterminancy one has to connect pairs of non-degenerate eigenvalues by cuts in the λ plane. The Wronskian becomes a hyperelliptic Riemann surface. The evolution of the unknown NSE solutions (called hyperelliptic functions $\mu_j(x,t)$) on this surface is as complicated as the original NSE equation. However, via the *Abel map* the Riemann surface is mapped onto a torus and on this torus the motion magically becomes *linear*. After obtaining the space-time dependence on the torus we need to come back to the original framework. This is called the Jacobi inversion and can be done exactly in terms of Riemann *theta* functions, giving at the end the exact solution of the NSE.

Not only the geometric approach is more clear but it also allows a treatment of the stability problem for the solutions of the NSE since it allows to trace the changes of the main spectrum after a perturbation of the initial condition. The following subsections provide a more detailed discussion of the IST of a periodic domain. The next section will discuss the stability. This information is available from the abundant literature on the IST and is only mentioned here in order to understand the mechanism governing the stability of the envelope of the ion wave turbulence. For this reason we will focus on the determination of the *main spectrum* and its role in the construction of the solution. The effective steps to be undertaken to obtain the solution will only be briefly described. We strongly recommend the lecture of Ref.[14] on which this presentation is based. In the following Sections we use the **generic notation** x for the space variable, and later return to the physically meaningful variables.

4.2 The Lax Operators and the Main Spectrum

The IST method starts by introducing a pair of linear operators, called the **Lax pair** (see [14]) which allow to express the nonlinear equation as a compatibility condition for a system of two linear equations. For the cubic Schrödinger equation

$$i\frac{\partial u}{\partial t} + \frac{\partial^2 u}{\partial x^2} + 2\left|u\right|^2 u = 0 \tag{100}$$

the **Lax pair** of operators is

$$L \equiv \begin{pmatrix} i\partial_x & u\left(x,t\right) \\ -u^*\left(x,t\right) & -i\partial_x \end{pmatrix} \tag{101}$$

$$A \equiv \begin{pmatrix} i\left|u\right|^2 - 2i\lambda^2 & -u_x + 2i\lambda u \\ u_x^* + 2i\lambda u^* & -i\left|u\right|^2 + 2i\lambda^2 \end{pmatrix} \tag{102}$$

The action of the operators on a two-component (column) wave function $\phi = \begin{pmatrix} \phi_1 \\ \phi_2 \end{pmatrix}$ is given by the equations

$$L\phi = \lambda\phi \tag{103}$$

$$\frac{\partial \phi}{\partial t} = A\phi \tag{104}$$

and the condition of compatibility of these two equations $\phi_{xt} = \phi_{tx}$ is precisely the cubic Nonlinear Schrödinger Equation.

4.3 Solving the Eigenvalue Equation for the Operator L

As in the "infinite-domain" IST, we have to solve the eigenvalue equation Eq.(103) using the initial condition for the function $u\left(x,t\right)$, $u\left(x,0\right)$. However, for the particular case of **periodic** solutions of the NSE, we must have $u\left(x+d,0\right) = u\left(x,0\right)$ where d is the length of the spatial period (*i.e.* actually L in the previous notation; however we use d in this and the next section, keeping L for the *Lax* operator). Fixing an arbitrary base point $x = x_0$ one considers the two independent solutions of the equation (103) which take the following "initial" values at $x = x_0$:

$$\phi\left(x_0\right) = \begin{pmatrix} 1 \\ 0 \end{pmatrix} \quad \text{and} \quad \widetilde{\phi}\left(x_0\right) = \begin{pmatrix} 0 \\ 1 \end{pmatrix} \tag{105}$$

The matrix $\Phi(x, x_0; \lambda)$ of solutions is given by

$$\Phi(x, x_0; \lambda) \equiv \begin{pmatrix} \phi_1(x, x_0; \lambda) & \widetilde{\phi}_1(x, x_0; \lambda) \\ \phi_2(x, x_0; \lambda) & \widetilde{\phi}_2(x, x_0; \lambda) \end{pmatrix} \tag{106}$$

i.e. this solution satisfies the equation

$$L\Phi = \lambda\Phi \text{ and } \Phi(x_0, x_0; \lambda) = \begin{pmatrix} 1 & 0 \\ 0 & 1 \end{pmatrix} \tag{107}$$

It can be seen that the Wronskian of $\left(\phi, \widetilde{\phi}\right)$ is the *determinant* of the matrix of fundamental solutions

$$W\left(\phi, \widetilde{\phi}\right) = \det(\Phi)$$

It can be shown that $\frac{\partial W}{\partial x} = 0$ *i.e.* the Wronskian is constant, which gives

$$\det(\Phi(x)) = \det(\Phi(x_0)) = 1$$

Using the initial condition for Φ one can solve the equation (107) and obtain $\Phi(x, x_0; \lambda)$; in particular we can obtain it at $x_0 + d$, *i.e.* after a period :

$$\Phi(x_0 + d, x_0; \lambda) = \begin{pmatrix} \phi_1(x_0 + d, x_0; \lambda) & \widetilde{\phi}_1(x_0 + d, x_0; \lambda) \\ \phi_2(x_0 + d, x_0; \lambda) & \widetilde{\phi}_2(x_0 + d, x_0; \lambda) \end{pmatrix} \tag{108}$$

This is called **the monodromy matrix** and is noted

$$M(x_0; \lambda) \equiv \Phi(x_0 + d, x_0; \lambda) \tag{109}$$

The monodromy matrix is the matrix of the fundamental solutions calculated for one spatial period. In general the *monodromy matrix* is the element of the monodromy group associated to a loop and to its homotopy class, for a marked point on a manifold (here the circle $S^1 \equiv [0, d)$. The linear monodromy group acts by replacing the elements of the vector column by *linear* combinations of the initial ones. For changes of the marked point x_0 the matrix has two invariants : **the trace** and **the determinant**:

$$[Tr\, M](x_0; \lambda) = [Tr\, M](\lambda) \equiv \Delta(\lambda) \tag{110}$$

$$\det M = 1 \tag{111}$$

The function $\Delta(\lambda)$ is called **the discriminant** and is independent of x_0.

Looking for Bloch (Floquet) solutions The values the discriminant $\Delta(\lambda)$ takes on the complex λ-plane control the monodromy and by consequence select the values of λ for which admissible eigenfunction of the Lax problem exist. In other words Δ governs the *spectral properties of the operator* L. To find under what conditions periodic solutions are possible, we construct the Bloch (or Floquet) solutions of the equation $L\phi = \lambda\phi$. The Bloch function has the property

$$\psi(x+d;\lambda) = e^{ip(\lambda)}\psi(x;\lambda) \tag{112}$$

where $p(\lambda)$ is the **Floquet exponent**. Like any other solution of the Lax eigenproblem ψ can be expressed as a linear combination of the two fundamental solutions ϕ and $\widetilde{\phi}$

$$\psi(x;\lambda) = A\phi(x;\lambda) + B\widetilde{\phi}(x;\lambda) \tag{113}$$

For the particular point x_0 we have, taking into account the "initial" conditions (105)

$$\psi(x_0;\lambda) = A\phi(x_0;\lambda) + B\widetilde{\phi}(x_0;\lambda) = \begin{pmatrix} A \\ B \end{pmatrix} \tag{114}$$

After one period d the function ψ is linearly modified by the monodromy matrix and, according to Eq.(112) is multiplied by a complex number of unit absolute value. We write $m(\lambda)$ for this number and look for those λ's where this is complex of modulus one. The regions of the λ plane where $m(\lambda)$ has modulus different of one correspond to unstable functions ψ. We write

$$\psi(x_0+d;\lambda) = m(\lambda)\,\psi(x_0;\lambda) \tag{115}$$

$$M\begin{pmatrix} A \\ B \end{pmatrix} = m(\lambda)\begin{pmatrix} A \\ B \end{pmatrix} \tag{116}$$

The Bloch functions are invariant directions for the *monodromy operator* M acting in the base formed by the fundamental solutions of the Lax eigenproblem. The monodromy operator preserves these directions and multiplies the Bloch vector with $m(\lambda) = \exp(ip(\lambda))$.

$$\det(M-m) = m^2 - (Tr\,M)\,m + \det M = m^2 - \Delta(\lambda)\,m + 1 = 0 \tag{117}$$

This gives

$$m^{\pm}(\lambda) = \frac{\Delta(\lambda) \pm \left(\Delta^2(\lambda) - 4\right)^{1/2}}{2} \tag{118}$$

In general $\Delta(\lambda)$ is an analytic function on the plane of the complex variable λ. The equation $\mathrm{Im}\left[\Delta(\lambda)\right] = 0$ is a single relation with two unknowns, $(\mathrm{Re}\lambda, \mathrm{Im}\lambda)$ and calculating one of them leaves the dependence on the other as a free parameter. This gives a curve in the plane and the real-λ axis is such a curve: for $\mathrm{Im}\lambda = 0$ we have $\mathrm{Im}\Delta(\lambda) = 0$.

We look at the variation of $\Delta(\lambda)$ on the complex plane to find the effect on $m^{\pm}(\lambda)$.

- Where the Floquet multiplier is a complex number of magnitude unity, the functions are only effected by a phase factor after one period. The Bloch functions (eigenfunctions of L) are stable under translations with d. The *real* λ axis is a region of stability. The region where Δ is real and
$$\Delta(\lambda) < 4 \tag{119}$$
is the **band of stability**, the modulus of m being 1.
- When λ is such that the discriminant ($\Delta(\lambda) \equiv Tr(M)$) is equal to 4, the eigenvalues of the monodromy matrix, $m(\lambda)$, are ± 1, and the Bloch functions are *periodic* or *antiperiodic*. **This is the Main Spectrum** in λ and is noted $\{\lambda_j\}$. The main spectrum consists of a set of discrete values λ_j.
- for those λ which gives $|m(\lambda)| \neq 1$ the Bloch eigenfunctions are unstable.

Consider the case where $\Delta(\lambda)$ is **complex**. For all values of the parameter λ on the complex plane, [with the exception of the Main Spectrum where $\Delta(\lambda) = \pm 2$], the eigenvalues of the monodromy matrix are distinct which means that the eigenfunctions Bloch are distinct and **independent** (The **Wronskian** is different of zero).

$$\begin{aligned} \psi^+(x+d) &= m^+\psi^+(x) \\ \psi^-(x+d) &= m^-\psi^-(x) \end{aligned} \tag{120}$$

However we can only be intersted in points of the *main spectrum*, as they can provide admissible (periodic) eigenfunctions of the Lax operator. (1) There are points λ_j in the main spectrum where the two eigenfunctions Bloch are distinct and independent: these points are called *degenerate* (with the meaning that for two independent eigenfunctions there is only one eigenvalue). (2) There are points λ_j where the two eigenfunctions Bloch are NOT independent: these points are called *nondegenerate*. For such values of λ the Wronskian is zero.

The important class of initial conditions $u(x, t = 0)$ with the property that there is only a *finite number of non-degenerate eigenvalues* λ_j is called **finite-band potential**.

4.4 The "Squared" Eigenfunctions

Starting from the initial condition $u(x, 0)$ and leting it evolve in time according to the NSE, $u(x, 0) \overset{NSE}{\rightarrow} u(x, t)$, the **discriminant** remains invariant. All the spectral structure obtained from the discriminant $\Delta(\lambda)$, *i.e.* the main spectrum, the stability bands, are invariant. The eigenvalues of the monodromy matrix λ_j are invariant.

4.4.1 The Two Bloch Functions for the Operator L

Consider the two Bloch eigenfunctions of the operator L:

$$\psi^+ = \begin{pmatrix} \psi_1^+ \\ \psi_2^+ \end{pmatrix} \text{ and } \psi^- = \begin{pmatrix} \psi_1^- \\ \psi_2^- \end{pmatrix}$$

and define the following functions

$$f(x,t;\lambda) \equiv -\frac{i}{2}\left(\psi_1^+\psi_2^- + \psi_2^+\psi_1^-\right) \quad (121)$$

$$g(x,t;\lambda) \equiv \psi_1^+\psi_1^- \quad (122)$$

$$h(x,t;\lambda) \equiv -\psi_2^+\psi_2^- \quad (123)$$

These functions have the property that they are periodic, $f(x+d,t;\lambda) = f(x,t;\lambda)$for all values of λ. Since f is composed of ψ^+ and ψ^- and these are Bloch solutions for the eigenvalue equation $L\varphi = \lambda\varphi$ we have

$$\frac{\partial f}{\partial x} = u^* g - uh \quad (124)$$

$$\frac{\partial g}{\partial x} = -2i\lambda g - 2uf \quad (125)$$

$$\frac{\partial h}{\partial x} = 2i\lambda h + 2u^* f \quad (126)$$

and similarly we write the time derivatives taking account of the compatibility condition which involves the operator A.

$$\frac{\partial f}{\partial t} = -\left(u_x^* + 2i\lambda u^*\right) g + i\left(-u_x + 2i\lambda u\right) h \quad (127)$$

$$\frac{\partial g}{\partial t} = 2i\left(|u|^2 - 2\lambda^2\right) g + 2i\left(-u_x + 2i\lambda u\right) f \quad (128)$$

$$\frac{\partial h}{\partial t} = -2i\left(|u|^2 - 2\lambda^2\right) h - 2i\left(-u_x^* + 2i\lambda u^*\right) f \quad (129)$$

It is shown in Ref.([14]) that the condition on the initial function $u(x,0)$ for there to be only a finite number of nondegenerate points in the main spectrum is equivalent to the requirement that f, g and h be finite-order polynomials in the parameter λ, which we take of degree $N+1$: $f(x,t;\lambda) = \sum_{j=0}^{N+1} f_j(x,t)\lambda^j$, $g(x,t;\lambda) = \sum_{j=0}^{N+1} g_j(x,t)\lambda^j$, $h(x,t;\lambda) = \sum_{j=0}^{N+1} h_j(x,t)\lambda^j$. From the Eqs.(124 - 126) it results however that g and h have degree N in λ.

One can check that the following combination is invariant in time and space and we note that it is actually the square of the Wronskian:

$$\frac{\partial}{\partial x}\left(f^2 - gh\right) = \frac{\partial}{\partial t}\left(f^2 - gh\right) = 0 \quad (130)$$

$$f^2 - gh = -\frac{1}{4}\left[W\left(\psi^+,\psi^-\right)\right]^2 \quad (131)$$

The Wronskian only depends on the spactral parameter λ. We also know that for a subset of the *main spectrum*, the nondegenerate eigenvalues, the Wronskian is zero. Then the number of the nondegenerate eigenvalues is $2N+2$ (the degree of $f^2 - gh$ as a polynomial in λ) and we express the function $f^2 - gh$ as a polinomial with constant coefficients (not depending on x and t).

$$-\frac{1}{4}\left[W\left(\psi^+,\psi^-\right)\right]^2 = f^2 - gh \equiv P(\lambda) = \sum_{k=1}^{2N+2} P_k\lambda^k = \prod_{j=0}^{2N+2}(\lambda-\lambda_j) \tag{132}$$

4.4.2 Product Expansion of g and Its Zeros $\mu_j(x,t)$. Introduction of the *Hyperelliptic* Functions μ_j (the Auxiliary Spectrum)

The functions g and h are both of order N in λ. For g we will note the N roots by $\mu_j(x,t)$. The coefficient of λ^{2N+2} in $f^2 - gh$ is f_{N+1}^2. Due to Eq.(130) it is a constant which can be taken 1. Now since $f_{N+1} = 1$ the coefficient of λ^N in g is $iu(x,t)$. Written as a product

$$g = iu(x,t)\prod_{j=1}^{N}\left[\lambda - \mu_j(x,t)\right] \tag{133}$$

By similar arguments we find that the coefficient of λ^N in h is $iu^*(x,t)$ and the zeros of h are $\mu_j^*(x,t)$. The functions $\mu_j(x,t)$ are called the *hyperelliptic functions*. It will be proved below that finding the *hyperelliptic* functions leads immediately to the solution $u(x,t)$.

Now we calculate the expression in Eq.(132) for $\lambda =$ a zero of the function g, *i.e.* λ is equal to the *hyperelliptic function* $\mu_j(x,t)$:

$$f^2(x,t;\lambda=\mu_m) - gh = f^2(x,t;\mu_m) = P(\lambda=\mu_m) \tag{134}$$

or

$$f(x,t;\mu_m) = \sigma_m\sqrt{P(\mu_m)} \tag{135}$$

Here the factor σ_m is a sheet index that indicates which sheet of the Riemann surface associated with $\sqrt{P(\lambda)}$ the complex μ_m lies on.

Let us calculate from Eq.(125) and Eq.(133) the derivative at x of μ_m:

$$\begin{aligned}\frac{\partial g(x,t;\lambda)}{\partial x} &= i\frac{\partial u(x,t;\lambda)}{\partial x}\prod_{j=1}^{N}(\lambda-\mu_j(x,t)) \\ &\quad -iu(x,t;\lambda)\sum_{j=1}^{N}\frac{\partial\mu_j(x,t)}{\partial x}\prod_{k=1,k\neq j}^{N}(\lambda-\mu_k(x,t))\end{aligned} \tag{136}$$

Replace here a zero of g: $\lambda = \mu_m$

$$\frac{\partial g(x,t;\mu_m)}{\partial x} = -iu(x,t;\mu_m)\frac{\partial\mu_m(x,t)}{\partial x}\prod_{k=1,k\neq m}^{N}(\mu_m-\mu_k(x,t)) \tag{137}$$

On the other hand we have, from Eq.(125)

$$\begin{aligned} \frac{\partial g}{\partial x} &= -2i\mu_m g\left(x,t;\lambda=\mu_m\right) - 2u\left(x,t;\lambda=\mu_m\right) f\left(x,t;\lambda=\mu_m\right) \\ &= -2uf \end{aligned} \tag{138}$$

then, since we have calculated $f\left(x,t;\lambda=\mu_m\right)$, Eq(135), we obtain

$$\frac{\partial \mu_m\left(x,t\right)}{\partial x} = \frac{-2i\sigma_m\sqrt{P\left(\mu_m\right)}}{\prod_{k=1,k\neq m}^{N}\left(\mu_m - \mu_k\left(x,t\right)\right)} \tag{139}$$

for $m = 1, 2, ..., N$.

The coefficients of the term with λ^N in the equation (125) are matched and we obtain $iu_x = -2ig_{N-1} - 2uf_N$.Using Eq.(133) we find

$$\partial_x \ln u = 2i\left(\sum_{j=1}^{N}\mu_j + f_N\right) \tag{140}$$

From the Eq.(132) we obtain the coefficient of the term λ^N, *i.e.* f_N

$$f_N = -\frac{1}{2}\sum_{k=1}^{2N+2}\lambda_k \tag{141}$$

Then from the preceding equation it results

$$\partial_x \ln u = 2i\left(\sum_{j=1}^{N}\mu_j - \frac{1}{2}\sum_{k=1}^{2N+2}\lambda_k\right) \tag{142}$$

The same procedure is applied to the equation for g_t, (128) and we obtain

$$\frac{\partial \mu_j\left(x,t\right)}{\partial t} = -2\left(\sum_{m\neq j}\mu_m - \frac{1}{2}\sum_{k=1}^{2N+2}\lambda_k\right)\frac{\partial \mu_j\left(x,t\right)}{\partial x} \tag{143}$$

$$\begin{aligned} \partial_t \ln u &= 2i\left[\sum_{j>k}\lambda_j\lambda_k - \frac{3}{4}\left(\sum_{k=1}^{2N+2}\lambda_k\right)^2\right] \\ &\quad -4i\left[\left(-\frac{1}{2}\sum_{k=1}^{2N+2}\lambda_k\right)\left(\sum_{j=1}^{N}\mu_j\right) + \sum_{j>k}\mu_j\mu_k\right] \end{aligned} \tag{144}$$

We can see that the problem has been reformulated: from the equation for the function $u\left(x,t\right)$ the Lax-operators formulation leads to the problem for Bloch functions, ψ^+ and

ψ^- and their Wronskian; then, using the "squared" eigenfunctions f, g and h we arrive at a formulation for the *hyperelliptic* functions $\mu_j(x,t)$. Finding $\mu_j(x,t)$ leads immediately to $u(x,t)$.

These operations have a significative geometric counterpart: every point of the complex plane of the spectral parameter λ is mapped, via Eq.(132) on the two-sheeted Riemann surface $\sqrt{-\frac{1}{4}W(x,t;\lambda)} = \sqrt{P(\lambda)}$ with singularities at the *nondegenerate* points of the main spectrum, λ_j. Removing the indeterminancy (by cutting and glueing the two sheets) we obtain a **hyperelliptic Riemann surface** of genus $g = N$. It results that we have to consider the variables $\mu_j(x,t)$, $j = 1, N$, as points on this surface but the motion of $\mu_j(x,t)$ is not simpler than the equation itself.

Now the next steps would be: (1) using the initial condition $\mu_j(0,0)$ on the function $\mu_j(x,t)$ we can solve the two equations (139) and (143) and find $\mu_j(x,t)$; (2) then using the initial condition $|u(0,0)|$ for the function $u(x,t)$ we can solve the equations (142) and (144). The procedure will be:

- Take the parameters $\{\lambda_j |\, j = 1, 2, ..., 2N+2\}$ as known; these parameters are **non-degenerate eigenvalues** from the main spectrum, they can be found from the equation $\Delta(\lambda) = \pm 2$ and Δ is determined from $u(x,0)$, the initial condition.
- Choose initial conditions $\mu_j(0,0)$ and $|u(0,0)|$ (see below)
- solve the equations for $\mu_j(x,t)$; this means to find the *hyperelliptic functions*. Then solve the equations for $u(x,t)$.

4.5 Solution: The Two-Sheeted Riemann Surface (Hyperelliptic Genus-N Riemann Surface)

The variable μ_j's are points lying on the two-sheeted Riemann surface associated with

$$\sqrt{P(\lambda)} = \left(\prod_{k=1}^{2N+2} (\lambda - \lambda_k)\right)^{1/2} \equiv R(\lambda) \tag{145}$$

which has **branch cuts at each of the nondegenerate points** λ_k. To eliminate the indeterminacy related to the square-root singularities (located at λ_j's) the surfaces must be cut and reglued, obtaining a complex manifold of complex dimension one, a hyperelliptic Riemann surface. We need some constructions on this surface.

4.5.1 Holomorphic Differential Forms and Cycles on the Riemann Surface

On this hyperelliptic Riemann surface (denoted M) it is possible to define N linearly independent holomorphic (regular) differentials. The following is the canonical basis of differentials one-forms defined on the manifold M:

$$dU_1 \equiv \frac{d\lambda}{R(\lambda)} \tag{146}$$

$$dU_2 \equiv \frac{\lambda \, d\lambda}{R(\lambda)}, \quad \cdots \tag{147}$$

$$dU_j \equiv \frac{\lambda^j \, d\lambda}{R(\lambda)}, \quad \cdots \tag{148}$$

$$dU_N \equiv \frac{\lambda^{N-1} \, d\lambda}{R(\lambda)} \tag{149}$$

On the surface M there are $2N$ topologically distinct closed curves (**cycles**). On a torus ($N = 1$, called *elliptic surface*) there are two curves which cannot be deformed one into another. A more general Riemann surface, with $N > 1$ can be shown to be equivalent to a sphere with N handles and is called *hyperelliptic Riemann surface*. The number N is called the *genus* of the surface. We have

$N =$ genus of the surface $=$ number of independent holomorphic differential forms

There are $2N$ cycles which are split into two classes: a_j cycles and b_j cycles. Each of these cycles has a specified direction (an arrow) attached to it. To construct the cycles, the rules to be applied are:

1. a_j cycles do not cross any other a_j cycle; b_j cycles do not cross any other b_j cycle;
2. the cycle a_k intersects b_k only once and intersects no other b cycle;
3. the intersection is such that, at the point of intersection, the vector tangent to the cycle a_k, the vector tangent to the cycle b_k and the normal to the tangent plane of M represent a system of three vectors compatible with the *orientation* of M.

This can be represented as:

$$a \circ a = 0 \ , \ b \circ b = 0 \ , \ a_j \circ b_k = \delta_{jk} \tag{150}$$

4.5.2 Periods and Matrices of Periods

With the set of holomorphic differential forms and the set of cycles we can define the **matrices of periods**, A and B

$$A_{kj} \equiv \int_{a_j} dU_k \tag{151}$$

$$B_{kj} = \int_{b_j} dU_k \tag{152}$$

and the matrices A and B are **invertible**. A change of basis of holomorphic differential forms is a linear transformation represented by the matrix C:

$$d\psi_j \equiv \sum_{k=1}^{N} C_{jk} \, dU_k \tag{153}$$

We can choose the matrix C as

$$C = A^{-1} \tag{154}$$

Then the periods in the new basis becomes

$$\int_{a_n} d\psi_j = \sum_{k=1}^{N} C_{jk} \int_{a_n} dU_k = \sum_{k=1}^{N} C_{jk} A_{kn} = \delta_{jn} \tag{155}$$

$$\int_{b_n} d\psi_j = \tau_{jn} \tag{156}$$

where

$$\tau = A^{-1} B \tag{157}$$

It can be shown that the matrix τ is symmetric $\tau_{jk} = \tau_{kj}$ and has a positive-definite imaginary part $\mathrm{Im}\tau > 0$.

4.5.3 The Abel Map

The Abel map is defined from the Riemann surface M to the space C^N and associates to a hyperelliptic Riemann surface (a sphere with N hadles) a N-torus. Using the differentials $d\psi_j$ one constructs a change of variables by the following procedure:

- choose a base point on M and call it p_0;
- define the variables $W_j(x,t)$ as integrals on the Riemann λ-surface of the *differential forms* from the arbitrary point p_0 to the point which is the **hyperelliptic function** $\mu_k(x,t)$:

$$\begin{aligned} W_j(x,t) &= \sum_{k=1}^{N} \int_{p_0}^{\mu_k(x,t)} d\psi_j \\ &= \sum_{k=1}^{N} \sum_{m=1}^{N} C_{jm} \int_{p_0}^{\mu_k(x,t)} \frac{\lambda^{m-1} d\lambda}{R(\lambda)} \end{aligned} \tag{158}$$

These variables have the remarkable property that their x and t dependence is trivial

$$\frac{d}{dx} W_j(x,t) = \sum_{k=1}^{N} \sum_{m=1}^{N} C_{jm} \frac{\mu_k^{m-1} \frac{d\mu_k}{dx}}{\sigma_k R(\mu_k)} \tag{159}$$

and using

$$\frac{d\mu_k(x,t)}{dx} = -2i\sigma_k \frac{R(\mu_k)}{\prod_{n \neq k} (\mu_k - \mu_n)} \tag{160}$$

it results

$$\frac{d}{dx}W_j(x,t) = \sum_{m=1}^{N} C_{jm} \sum_{k=1}^{N} \frac{-2i\mu_k^{m-1}}{\prod_{n\neq k}(\mu_k - \mu_n)} \tag{161}$$

To calculate the sum

$$\sum_{k=1}^{N} \frac{\mu_k^{m-1}}{\prod_{n\neq k}(\mu_k - \mu_n)} \tag{162}$$

one can use the contour integral

$$I_m \equiv \frac{1}{2\pi i}\int_C \frac{\lambda^{m-1}d\lambda}{\prod_{n\neq k}(\mu_k - \mu_n)}$$

where the contour C encloses all of the poles μ_n counterclockwise. By the residue theorem

$$\sum_{k=1}^{N} \frac{\mu_k^{m-1}}{\prod_{n\neq k}(\mu_k - \mu_n)} = I_m \tag{163}$$

One can evaluate this integral by compactifying the λ plane into a sphere and noticing that the contour encloses the pole at $z = \infty$. Evaluating the residuu at $z = \infty$

$$\sum_{k=1}^{N} \frac{\mu_k^{m-1}}{\prod_{n\neq k}(\mu_k - \mu_n)} = \delta_{m,N} \tag{164}$$

and from this we obtain

$$\frac{d}{dx}W_j(x,t) = -2iC_{j,N} \equiv \frac{1}{2\pi}k_j \tag{165}$$

A similar calculation gives

$$\begin{aligned} \frac{d}{dt}W_j(x,t) &= -4i\left[C_{j,N-1} + \left(\frac{1}{2}\sum_{k=1}^{2N+2}\lambda_k\right)C_{j,N}\right] \\ &= \frac{1}{2\pi}\Omega_j \end{aligned} \tag{166}$$

It results from these calculations

$$W_j(x,t) = \frac{1}{2\pi}(k_j x + \Omega_j t + d_j) \tag{167}$$

where d_j is a phase which is determined by the initial condition on μ_k.

The Abel map has *linearized* the motion of the points $\mu_k(x,t)$. Now, in order to determine the function $u(x,t)$ we must invert the Abel map, returning from the variables W to μ. This is the *Jacobi inversion problem*.

4.5.4 The Jacobi Inversion Problem

The return from the variables $W_j(x,t)$ of the N-torus back to the variables $\mu_k(x,t)$ of the Riemann surface M (notice we will only need particular combinations of the functions μ) can be done in an exact way using the Riemann θ function.

The argument of θ is the N-tuple of complex numbers $\overline{z} \in C^N$.

$$\theta(z|\tau) \equiv \sum_{m_1=-\infty}^{\infty} \cdots \sum_{m_N=-\infty}^{\infty} \exp\left(2\pi i \overline{m} \cdot \overline{z} + \pi i \overline{m} \cdot \tau \cdot \overline{m}\right) \tag{168}$$

with the notations $\overline{m} \cdot \overline{z} = \sum_{k=1}^{N} m_k z_k$ and $\overline{m} \cdot \tau \cdot \overline{m} = \sum_{k,j=1}^{N} \tau_{jk} m_j m_k$. Adding to the vector z a vector with a single nonzero element in the position k we obtain from the definition:

$$\theta(z + e_k|\tau) = \theta(z|\tau) \tag{169}$$

which means that the θ -function has N real periods. Adding the full k-th column of the matrix τ, we obtain:

$$\theta(z + \tau_k|\tau) = \exp(-2\pi i z_k - \pi i \tau_{kk})\, \theta(z|\tau) \tag{170}$$

where τ_{kk} is the diagonal element of the $\underline{\tau}$ matrix.

There are two quantities which must be determined in order to have the explicit form of the solution. The space and respectively time equations for $u(x,t)$ depend on the quantity $\sum_{j>k} \mu_j \mu_k = \frac{1}{2}\left[\left(\sum_{m=1}^{N} \mu_m\right)^2 - \sum_{m=1}^{N} \mu_m^2\right]$. So we must determine the following combinations of the variables μ_j:

$$\sum_{m=1}^{N} \mu_m(x,t) \quad \text{and} \quad \sum_{m=1}^{N} \mu_m^2(x,t) \tag{171}$$

In order to calculate these two quantities, we shall start by introducing a series of functions , $\psi_j(p)$ defined on the Riemann surface M:

$$\psi_j(p) = \int_{p_0}^{p} d\psi_j \tag{172}$$

where p is a point on M , $d\psi_j$ is the j -th holomorphic differential form defined on M and p_0 is an arbitrary fixed point on M. The functions $\psi_j(p)$ are multiple valued since the contour is defined up to addition of any combination of the cycles on M.

Now consider the function, $F(p)$:

$$F(p) \equiv \theta(\psi(p) - K|\tau) \tag{173}$$

where θ is the N-dimensional Riemann *theta* function associated with the surface M; $\psi(p)$ is the N-dimensional column of complex functions $\psi_j(p)$; K is an N-dimensional column of complex numbers independent of the point p. The function $F(p)$ is multivalued on the Riemann surface M for the same reason as ψ: moving the point p on the surface such as to turn around one of the b cycle and returning to the initial position, the functions ψ_j will add elements of the τ matrix. This will make the function F to change as imposed by the properties of the θ function, shown in Eq.(170).

In order to render the function F single-valued, we replace its domain M by a new surface, obtained from M by dissecting it in a canonical fashion, along a basis of cycles. This new surface, M^* is simply connected and has the bord composed of a number of arcs equal to $4N$.

This operation is necessary in order to render the function $F(p)$ entire and allow us define the integral

$$I_0 = \frac{1}{2\pi i} \int_{\partial M^*} d \ln F(p) \tag{174}$$

around the contour of the surface M^*. Applying Cauchy theorem, the integral is the *number of zeroes* of $F(p)$ on the surface M^*. This number is N (Riemann).

We now impose the condition that the N zeros of $F(p)$ coincide with the N points $(\mu_j(x,t), \sigma_j)$. This fixes the values of the N complex numbers K_j $(j = 1, ..., N)$. By doing so the contour integral (174) calculated with the reziduum theorem will involve the variables μ_j.

Calculation of the Two Combination of μ's We must remember that the complex λ plane is covered by the two-sheeted Riemann surface whose compact version is M. Further this is mapped by Abel map onto the Jacobi N-torus. To a variable on the Riemann surface M (formally also on M^*), say p, corresponds a certain $\lambda(p)$. One defines the following integrals, which are proved to be real constants :

$$I_1 = \frac{1}{2\pi i} \int_{\partial M^*} \lambda(p)\, d \ln [F(p)] \equiv A_1 \tag{175}$$

$$I_2 = \frac{1}{2\pi i} \int_{\partial M^*} \lambda^2(p)\, d \ln [F(p)] \equiv A_2 \tag{176}$$

They can be evaluated by the residuu theorem. By the choice of the constants K's, the zeroes of the of the function $F(p)$ are located at μ_j. Then the residues are just the integrand (λ) calculated in μ plus the residuu at the infinite, $\pm\infty$:

$$I_1 = \sum_{m=1}^{N} \mu_m + \operatorname*{Res}_{\lambda\to\infty^+} [\lambda(p)\, d \ln F(p)] + \operatorname*{Res}_{\lambda\to\infty^-} [\lambda(p)\, d \ln F(p)] \tag{177}$$

$$I_2 = \sum_{m=1}^{N} \mu_m^2 + \operatorname*{Res}_{\lambda\to\infty^+} [\lambda^2(p)\, d \ln F(p)] + \operatorname*{Res}_{\lambda\to\infty^-} [\lambda^2(p)\, d \ln F(p)] \tag{178}$$

The reason to write **two residues** at infinity is that $\lambda = \infty$ is not a branching point which means that there are **two points** on the manifold M corresponding to $\lambda = \infty$, one on each sheet of the surface. We note $r^{\pm}$ the value of $\psi(p)$ when p is the point on M^* corresponding to $\lambda \to \infty^{\pm}$.

The result is

$$A_1 = \sum_{j=1}^{N} \mu_j + \frac{i}{2}\frac{\partial}{\partial x} \ln \left[\frac{F(r^- - K)}{F(r^+ - K)}\right] \tag{179}$$

or

$$A_1 = \sum_{j=1}^{N} \mu_j + \frac{i}{2}\frac{\partial}{\partial x} \ln \left[\frac{F(r^- - K)}{F(r^+ - K)}\right] \tag{180}$$

with $\theta^{\pm} = F(r^{\pm} - K)$. The expression of I_2

$$A_2 = \sum_{j=1}^{N} \mu_j^2 - \frac{1}{4}\frac{\partial}{\partial x} \ln \left[F(r^+ - K) F(r^- - K)\right] + \frac{i}{4}\frac{\partial}{\partial t} \ln \left[\frac{F(r^- - K)}{F(r^+ - K)}\right] \tag{181}$$

Return to the equations for the function u With these expressions we come back to the two equations for the two partial derivatives of u.

$$\frac{\partial}{\partial x} \ln u = \frac{\partial}{\partial x} \ln \frac{F(r^+ - K)}{F(r^- - K)} + 2iA_1 - i \sum_{m=1}^{2N+2} \lambda_m \tag{182}$$

$$\frac{\partial}{\partial t} \ln u = \frac{\partial}{\partial t} \ln \frac{F(r^+ - K)}{F(r^- - K)} + i\text{const} \tag{183}$$

The Solution Let us note

$$\omega_0 = Q \tag{184}$$

$$k_0 = 2A_1 - \sum_{m=1}^{2N+2} \lambda_m \tag{185}$$

which are "external" frequency and wavelength.

The solution is

$$u(x,t) = u(0,0) \exp(ik_0 - i\omega_0 t) \frac{\theta(W^-|\tau)}{\theta(W^+|\tau)} \tag{186}$$

where

$$W_j^{\pm} = \frac{1}{2\pi}\left(k_j x + \Omega_j t + \delta_j^{\pm}\right) \tag{187}$$

The phases $\delta_j^{\pm}$ are the part of $r^{\pm} - K_j$ which is independent of (x,t).

5 Stability of the Envelope Solutions

In this Section we return to the equations obtained by multiple space and time scales analysis. Then the meaning of the variables ϕ, y, t is: first order (envelope) amplitude of the potential fluctuations in the ion turbulence and respectively "large" (poloidal) space and "slow" time variables. We focus on a single NSE in order to examine the properties of the stability of the solutions. The equation has the generic form

$$i\frac{\partial\phi}{\partial t}+\alpha\frac{\partial^2\phi}{\partial y^2}+2\sigma\left|\phi\right|^2\phi=0 \tag{188}$$

and represents the equation for the envelope of a fast oscillation arising as solution of the barotropic equation. On infinite spatial domain this equation has soliton as solutions. On a periodic spatial domain (which is our case) the solutions can be cnoidal waves and they may be modulationally unstable. The instability can be examined in the neighbourhood of a solution by imposing a slight deviation and studying its time behaviour by classical perturbation expansion. This analysis and comparison with the geometric-algebraic results have been carried out in Ref.[14].

Consider $\phi\left(y,t\right)$ is a solution and perturb it at the initial moment $t=0$ as

$$\widetilde{\phi}\left(y,t\right)=\phi\left(y,0\right)+\varepsilon h\left(y\right)$$

The question is to find the relative behaviour of the two functions $\phi\left(y,t\right)$ and $\widetilde{\phi}\left(y,t\right)$. It they depart exponentially for time close to the initial moment, then $\phi\left(y,t\right)$ is an unstable solution. Suppose the initial solution $\phi\left(y,t\right)$ is the envelope of an exact plane wave *i.e.* it is constant in space. As a function of time it must have the form $\phi\left(y,t\right)=p\exp\left(2iqt\right)$ and we find

$$\phi\left(y,t\right)=p\exp\left(2i\sigma p^2t\right) \tag{189}$$

To this solution we add a perturbation

$$\widetilde{\phi}\left(y,t\right)=p\exp\left(2i\sigma p^2t\right)\left\{1+\varepsilon\left[A_1\exp\left(iky-i\Omega t\right)+A_2\exp\left(-iky+i\Omega t\right)\right]\right\} \tag{190}$$

where A_1 and A_2 are *real* coefficients. Linearising the equation for small ε we obtain:

$$\begin{aligned}&\left\{\left(2p^3\sigma+p\Omega-p\alpha k^2\right)\exp\left(iky-i\Omega t\right)+2p^3\sigma\exp\left(-iky+i\Omega t\right)\right\}A_1+\\&\left\{\left(2p^3\sigma-p\Omega^*-p\alpha k^2\right)\exp\left(-iky+i\Omega t\right)+2p^3\sigma\exp\left(iky-i\Omega t\right)\right\}A_2\\=&\ 0\end{aligned}$$

Taking the complex conjugate we obtain another equation for the two *real* coefficients A_1 and A_2. The condition of compatibility of this system of equations (the determinant equals zero) can only be fulfilled if the factor multiplying the function of y and t is allways zero. This gives us the dispersion relation

$$\Omega=\pm\sqrt{\alpha}k\left(\alpha k^2-4p^2\sigma\right)^{1/2} \tag{191}$$

which shows that for

$$|k| < 2|p|\sqrt{\frac{\sigma}{\alpha}} \tag{192}$$

(*i.e.* for long wavelengths) the two solutions $\phi(y,t)$ and $\widetilde{\phi}(y,t)$ diverge exponentially in time and the function $\phi(y,t)$ is modulationally unstable. From Eq.(191) we notice that when σ (the Landau coefficient)< 0 and α (the dispersion coefficient)> 0 the frequency Ω is real and there is no linear instability. This remark is useful in connection with the numerical calculations described above, which connect the physical parameters with the coefficients (α, σ) of the NSE. Sampling the space of the physical parameters we must retain those points where the linear instability is possible. In the above formula σ , α and p are dimensionless constants and the variables of the equation are normalised : time to t_0 and length to L. The value of p is connected with the frequency of oscillation of the uniform poloidal envelope ω_{unif}^{phys} (s^{-1}) by

$$p = \sqrt{\frac{t_0}{2\sigma}\omega_{unif}^{phys}} \tag{193}$$

and the condition of linear instability is for the poloidal wavenumber $\overline{m}$

$$\overline{m} < \sqrt{2}\frac{a}{L}\sqrt{\frac{\sigma}{\alpha}}\left(t_0\omega_{unif}^{phys}\right)^{1/2} \tag{194}$$

In the case of the following parameters: $\alpha = 0.399$, $\sigma = 1.452$, $d = 2\pi a/L = 2\pi(1/0.3) = 20.944$, $\omega_{unif}^{phys} = 10^5$ (sec^{-1}) we have

$$\overline{m} \lesssim 8 \tag{195}$$

For the Nonlinear Schrodinger Equation it is possible to follow the perturbed solution much further in time in an exact manner. In Ref.[14] it is shown that from a solution it is possible to return toward the initial time and to recover the results of the linear analysis. This is possible because the equation can be exactly solved for initial conditions starting arbitrary close to the envelope of the plane wave. The case of the plane wave (uniform envelope) is particular due to the simple form of the main spectrum, as will be discussed below.

5.1 Modulation Instability of the Envelope of a Plane Wave

5.1.1 General Method

The geometric-algebraic setting of Inverse Scattering Transform is the appropriate framework for understanding the nature of the stability of the exact solutions of the NSE. This is because the solution is strongly dependent on the topology of the hyperelliptic Riemann curve which in turn is dependent on the initial function. A perturbation of the initial condition may change the main spectrum turning degenerate eigenvalues into new nondegenerate pairs of eigenvalues. This means that the hyperelliptic curve will have a different genus $N' > N$ and new cycles and differential forms should be defined. The solution depends precisely on this topology.

We order to investigate the stability of an exact N-band solution of the NSE we have to develop in detail the construction of solution. In the particular case where the solution is the plane wave this analysis is simpler since all the spectral properties can be found analytically. It has been proved [14] that under the perturbation of the initial condition the degeneracies in the main spectrum will be removed and new degrees of freedom will become active. In the new solution some of the new degrees of freedom have little effect and can be neglected.

5.1.2 Calculation of the Main Spectrum Associated to the Uniform Amplitude Solution

The unperturbed solution is the envelope of the plane wave, Eq.(189). To deteremine the main spectrum, the solution is considered initial condition (at $t = 0$) and introduced in the Lax operator. The eigenvalue problem of the Lax operator has the form

$$\begin{pmatrix} i\sqrt{\frac{\sigma}{\alpha}}\frac{\partial}{\partial y} & 1 \\ -1 & -i\sqrt{\frac{\sigma}{\alpha}}\frac{\partial}{\partial y} \end{pmatrix} \Phi = \lambda\Phi \tag{196}$$

where

$$\Phi \equiv \begin{pmatrix} \phi_1 \\ \phi_2 \end{pmatrix} \tag{197}$$

The equation is:

$$\frac{\partial^2 \phi_1}{\partial y^2} + \frac{\alpha}{\sigma}\left(1+\lambda^2\right)\phi_1 = 0 \tag{198}$$

The general solution is written

$$\phi_1 = A\cos(\kappa y) + B\sin(\kappa y) \tag{199}$$

$$\kappa = \sqrt{\frac{\alpha}{\sigma}}\sqrt{1+\lambda^2} \tag{200}$$

The second component will be also a combination of harmonic functions:

$$\phi_2 = C\cos(\kappa y) + D\sin(\kappa y) \tag{201}$$

The equation can have two independent one-column solutions. The one corresponding to the boundary conditions

$$\Phi_1(y=0;\lambda) \equiv \left.\begin{pmatrix} \phi_1 \\ \phi_2 \end{pmatrix}\right|_{y=0} = \begin{pmatrix} 1 \\ 0 \end{pmatrix} \tag{202}$$

is determined by $A = 1$, $B = \sqrt{\alpha/\sigma}\,\lambda/(i\kappa)$, $C = 0$ and $D = \sqrt{\alpha/\sigma}\,i/\kappa$. The solution is

$$\Phi_1(y;\lambda) = \begin{pmatrix} \cos\kappa y - \lambda i\sqrt{\frac{\alpha}{\sigma}}\frac{\sin\kappa y}{\kappa} \\ i\sqrt{\frac{\alpha}{\sigma}}\frac{\sin\kappa y}{\kappa} \end{pmatrix} \tag{203}$$

The second fundamental solution corresponds to the initial condition;

$$\Phi_2 \left(y = 0; \lambda \right) = \begin{pmatrix} 0 \\ 1 \end{pmatrix} \tag{204}$$

and it results:

$$\Phi_2 \left(y; \lambda \right) = \begin{pmatrix} i\sqrt{\frac{\alpha}{\sigma}} \frac{\sin \kappa y}{\kappa} \\ \cos \kappa y + \lambda i \sqrt{\frac{\alpha}{\sigma}} \frac{\sin \kappa y}{\kappa} \end{pmatrix} \tag{205}$$

The fundamental solution matrix Φ is a 2×2 matrix having as column the two above solutions. From the fundamental solution matrix we calculate the monodromy matrix

$$M \left(\lambda \right) \equiv \Phi \left(y = d; \lambda \right) \tag{206}$$

The discriminant is the trace of the monodromy matrix.

$$\Delta \left(\lambda \right) = 2 \cos \left(d \sqrt{\frac{\alpha}{\sigma}} \sqrt{1 + \lambda^2} \right) \tag{207}$$

We can proceed to the calculation of the *spectrum.*

5.2 The Spectrum of the Plane Wave Solution

5.2.1 The Spectrum of the Unperturbed Initial Condition (Case of Plane Wave)

We take a set of constants as derived from the numerical calculations that connects the physical conditions to the coefficients of the NSE, $d = 2\pi \left(\frac{a}{L} \right)$, $\alpha = 0.399$, $\sigma = 1.452$. With these constants we calculate the spectrum on the complex λ-plane. Although we can have analytical solutions the graphs below show the difficulty of resolving the details of the variation of the functions. To be noted that the function is intersected with a plane arbitrarly placed at the magnitude $TrM = 5$, for easier 3d-representation.

This graph shows the presence of roots on the imaginary axis of λ of the equation $|TrM \left(\lambda \right) - 2| = 0$ with very narrow variation of the function. Actually it is easily to see that for the parameters adopted here, there are *three* roots of this equation, while in the graph we can hardly see two of them. A graphical better way of finding these roots is to look for the intersections of the contours of the two equations:

$$\begin{aligned} \mathrm{Re} \left(TrM \right) - 2 &= 0 \\ \mathrm{Im} \left(TrM \right) &= 0 \end{aligned} \tag{208}$$

where, again, the other two roots cannot be easily resolved. However the analytical expression allows to find the positions of the roots both on the Reλ axis and on the Imλ axis:

By contrast, the variation along the real λ axis easily evidences the zeroes of the corresponding to the *main spectrum.*

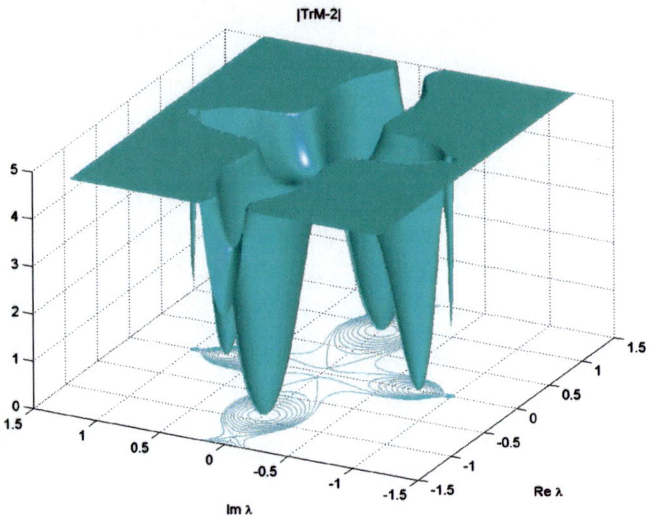

Figure 14: Absolute value of $Tr(M)-2$. The zeroes on the $\mathrm{Im}\lambda$ axis are difficult to resolve.

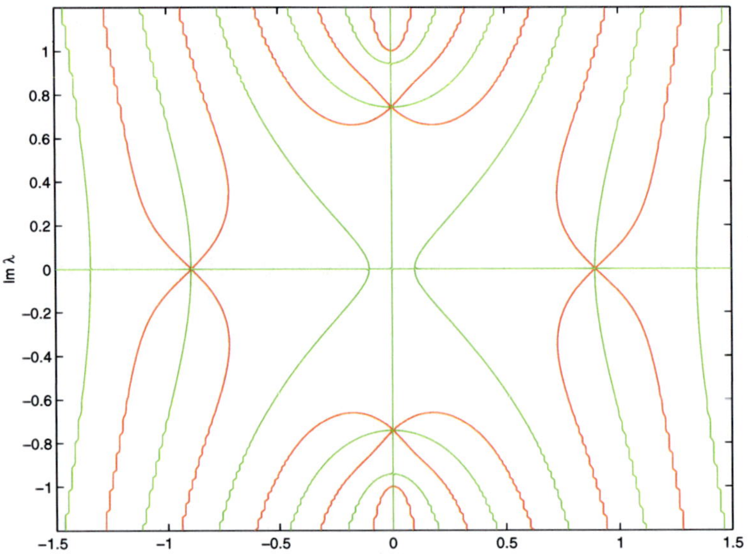

Figure 15: Contours of the solutions of: $TrM - 2 = 0$ and $\mathrm{Im}TrM = 0$.

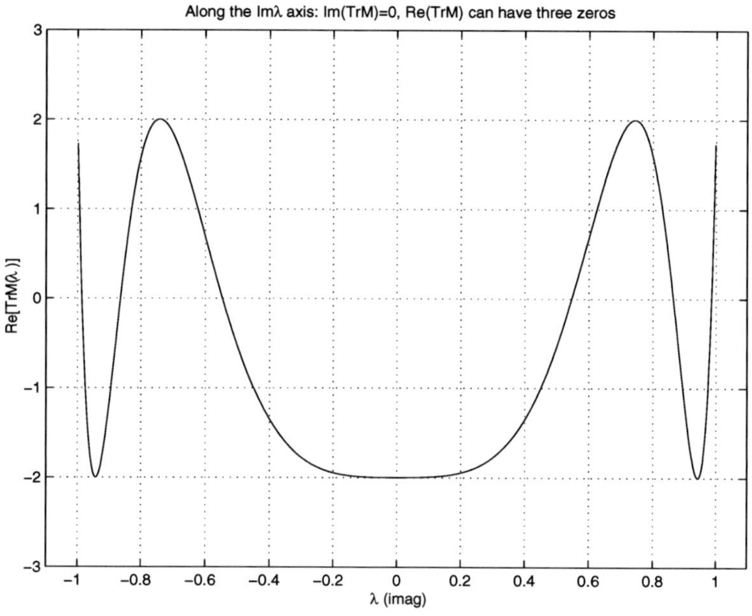

Figure 16: The function $\mathrm{Im}TrM = 0$ along the imaginary λ axis.

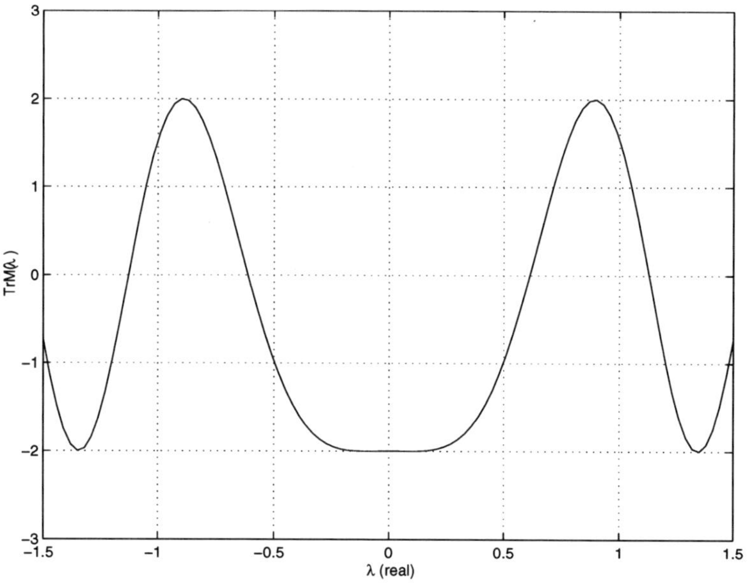

Figure 17: The function $\mathrm{Im}TrM = 0$ along the real λ axis.

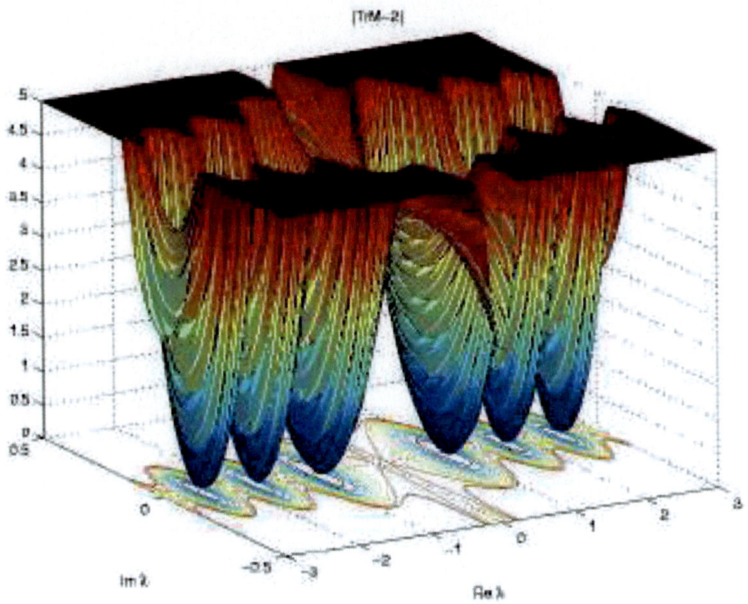

Figure 18: The graph of $Tr(M) - 2$ showing the zeroes on the Reλ axis.

The stability or instability of the Bloch fuctions (*i.e.* the values of the quantity $m(\lambda)$ which is the eigenvalue of the monodromy matrix) is governed by the *discriminant* $\Delta(\lambda)$. We have $-2 \leqslant \Delta(\lambda) \leqslant 2$ for all real λ and for $\lambda = i\alpha$ with $-1 \leqslant \alpha \leqslant 1$. The *main spectrum* corresponds to the values of λ for which $\Delta(\lambda) = \pm 2$. For *real* λ, $\lambda \equiv \lambda^0_{n,\text{real}}$ this means

$$\lambda^0_{n,\text{real}} = \pm \left[\left(\frac{n\pi}{d\sqrt{\alpha/\sigma}} \right)^2 - 1 \right]^{1/2} \tag{209}$$

for $n \geq 3$. For pure imaginary λ, $\lambda \equiv i\lambda^0_{n,imag}$

$$\lambda^0_{n,imag} = \pm \left[1 - \left(\frac{n\pi}{d\sqrt{\alpha/\sigma}} \right)^2 \right]^{1/2} \tag{210}$$

for $n = 0, 1, 2$ (with the values of the parameters α, σ and d). These values are: $i\lambda^0_{0,imag} = \pm i$, $i\lambda^0_{1,imag} = \pm i0.9422$ and $i\lambda^0_{2,imag} = \pm i0.7424$. Since d is finite only a finite number of degenerate pairs will appear on the imaginary axis. At $n = 0$ there are two eigenvalues $\lambda_0 = \pm i$. These eigenvalues are *nondegenerate* in the sense that there is only one Bloch function for each of these values of λ. This can be shown directly, constructing the Bloch functions form the matrix of fundamental solutions (203) and (205) with $\kappa \to 0$ (because $\Delta(\lambda) = 2$ implies $\kappa = 0$) and showing that only **one Bloch function** can be obtained.

5.2.2 The Spectrum of the Perturbed Initial Condition (Case of Plane Wave)

Starting from an initial condition slightly different of the *plane wave's envelope*

$$u(y,0) \to u(y,0) + \varepsilon h(y) \tag{211}$$

we shall have to repeat all steps in the determination of the spectrum. This must be done in a comparative manner, identifying the degenerate eigenvalues λ_j of the main spectrum which becomes nondegenerate after perturbation. The time behaviour of the exact solution developing from the perturbed initial condition is dominated by the imaginary part of the variables W which are obtained on the basis of the new (perturbed) quantities μ_j and this can be calculated.

Using the new initial condition Eq.(211) we shall calculate the new main spectrum, i.e. the new values of the parameter λ for which the discriminant is ± 2. All the spectrum of the simple initial condition $u(y,0)$ is perturbed by order-ε quantities. The main effect of this perturbation is that the degeneracies are broken. In order to construct the exact solution corresponding to the new position of the nondegenerate λ values, we need to specify the cycles on the two-sheeted Riemann surface. The steps are as usual and consists of first choosing the cycles on the Riemann surface and the base of the holomorphic differentials; then, compute the matrices of periods (the matrix of a periods, and the matrix of the b periods); compute the τ matrix; construct the Abel map, which maps the points of the compact two-sheeted Riemann surface onto the Jacobi torus; find the wavenumbers and the frequencies in the Jacobi manifold; make the Jacobi inversion, using the theta functions;

Finding the Base of the Holomorphic Differentials One of the eigenvalue pair of the unperturbed system is $(-i, i)$. Consider now that under the perturbation a degenerate eigenvalue λ_j^0 splits into the conjugated pair $\left(\lambda_k^0 - \varepsilon_k, \lambda_k^0 + \varepsilon_k\right)$. In Ref. [14] it was constructed a new basis as a linear combination of the canonical basis:

$$dU_j = \frac{1}{2\pi i}\sqrt{1+\left(\lambda_j^0\right)^2}\frac{\prod_{m\neq j}\left(\lambda - \lambda_m^0\right) d\lambda}{R(\lambda)}, \quad \text{for } j = 1, 2, ..., N \tag{212}$$

where

$$R(\lambda) = \sqrt{1+\lambda^2}\prod_{k=1}^{N}\left[\left(\lambda - \lambda_k^0 - \varepsilon_k\right)\left(\lambda - \lambda_k^0 + \varepsilon_k\right)\right]^{1/2} \tag{213}$$

To understand the choice of basis it is useful to look at the limit $\varepsilon \to 0$ where the basis of holomorphic differential becomes

$$\lim_{\varepsilon\to 0} dU_j = \frac{1}{2\pi i}\sqrt{1+\left(\lambda_j^0\right)^2}\frac{1}{\sqrt{1+\lambda^2}}\frac{d\lambda}{\left(\lambda - \lambda_j^0\right)} \tag{214}$$

The loop (cycle) a_j encircles the pole λ_j^0 and the period is

$$\lim_{\varepsilon\to 0} A_{kj} = \lim_{\varepsilon\to 0}\int_{a_j} dU_k = \int_{a_j}\lim_{\varepsilon\to 0} dU_k = \delta_{kj} \tag{215}$$

This is because the cycle a_j surrounds the pole λ_j^0 or it does not surrounds any other pole.

We find that, under the *limit operation* $\varepsilon \to 0$, the matrix $\mathbf{A}$ goes over to the identity matrix to the first order in ε, $O(\varepsilon)$ and that the matrix $\mathbf{B}$ becomes

$$\lim_{\varepsilon \to 0} B_{jk} = \tau_{jk} + O(\varepsilon) \tag{216}$$

To see what happens with the $\mathbf{B}$ matrix in this limit, we must examine the *off-diagonal* terms and the diagonal terms.

The *off-diagonal* terms are well-behaved and can be integrated analytically:

$$\lim_{\varepsilon \to 0} B_{jk} = \int_{b_j} \lim_{\varepsilon \to 0} dU_k \text{ for } j \neq k \tag{217}$$

and

$$\tau_{kj} = \frac{1}{2\pi i}\sqrt{1 + \left(\lambda_j^0\right)^2} \int_{b_j} \frac{1}{\sqrt{1+\lambda^2}} \frac{d\lambda}{\left(\lambda - \lambda_k^0\right)} + O(\varepsilon) \tag{218}$$

The contour b_j passes through the middle of the segment connecting the two new λ_j and goes around the pole at $\lambda = -i$, for example. If this contour does not surround the pole at λ_k^0 then the integral is elementary. If the contour intersecting the cut from $-i$ to infinity encircles the pole at λ_k^0 then it may be changed such that to intersect the other cut, connecting the branch point $+i$ to infinity, and so it does not encircle any pole. In conclusion, since the off-diagonal elements of the matrix τ depend (to $0(\varepsilon)$) only of the positions of the original double points, which are determined only by the parameter d, it results that: **the off-diagonal terms of the τ matrix does not carry any information on the** *perturbed initial condition.* **Only the diagonal terms of the τ matrix are affected by the initial condition.**

The *diagonal* terms of the τ - matrix are *singular* at the limit $\varepsilon \to 0$. Consider

$$\tau_{kk} = \frac{1}{2\pi i}\sqrt{1 + \left(\lambda_j^0\right)^2} \int_{b_k} \frac{1}{\sqrt{1+\lambda^2}} \frac{d\lambda}{\sqrt{\left(\lambda - \lambda_k^0\right)^2 - \varepsilon_k^2}} + O(\varepsilon) \tag{219}$$

Suppose that

λ_k^0 is on the **imaginary axis**

Then we have

$$\tau_{kk} = \frac{1}{2\pi i}\sqrt{1 + \left(\lambda_j^0\right)^2} \left(2\int_{-i}^{\lambda_k^0} + \frac{1}{2}\int_{a_k}\right) \frac{1}{\sqrt{1+\lambda^2}} \frac{d\lambda}{\sqrt{\left(\lambda - \lambda_k^0\right)^2 - \varepsilon_k^2}} + O(\varepsilon) \tag{220}$$

$$\tau_{kk} = \frac{1}{2} + \frac{1}{\pi i}\sqrt{1 + \left(\lambda_j^0\right)^2} \int_{-i}^{\lambda_k^0} \frac{1}{\sqrt{1+\lambda^2}} \frac{d\lambda}{\sqrt{\left(\lambda - \lambda_k^0\right)^2 - \varepsilon_k^2}} + O(\varepsilon) \tag{221}$$

The integral can be made with *elliptic functions* but an approximation is (near $\varepsilon_k = 0$)

$$\tau_{kk} = \frac{1}{2} - \frac{i}{\pi} \ln |\varepsilon_k| + O(1) + O(\varepsilon_k) \tag{222}$$

Consider now

λ_k^0 is on **real axis**

then the integral defining the diagonal elements of the matrix τ (the integral along the b_k cycles) is

$$\int_{-i}^{\lambda_k^0} = \int_{-i}^{0} + \int_{0}^{\lambda_k^0} \tag{223}$$

where the second integral is along the **real axis**.

The first integral is done with the residuu theorem after taking the limit $\varepsilon_k = 0$ and gives a finite imaginary contribution which is well-behaved for this limit. The second can be done and the result is

$$\tau_{kk} = -\frac{i}{\pi} \ln |\varepsilon_k| + iO(1) + O(\varepsilon_k) \tag{224}$$

We then conclude that for $\varepsilon_k \to 0$ the b - period matrix τ becomes logarithmically singular on the main diagonal.

The main conclusion is: (1) The real-λ axis degenaracies have Ω_j real so they are linearly stable. The imaginary axis degeneracies have Ω_j imaginary so they are linearly unstable.

This conclusion can have important practical effects: in our case the degenerate eigenvalues on the *imaginary* λ-axis split and induce instability of the plane wave solution, *i.e.* of the envelope. At the same time the solution evolves toward one of the stable (soliton-like) solutions. Still much work is required to examine the stability of that solution to additional perturbation. For example, the main spectrum of the solution of the type $\sec h(y)$ can be obtained from the discriminant whose form is very complicated, with many purely imaginary (dangerous) eigenvalues.

6 The Growth of the Perturbed Solution as a Source of Poloidal Asymmetry and Spontaneous Plasma Spin-Up

6.1 The Minimal Condition for Instability of the Poloidal Rotation

We have found that the space-uniform solution (poloidal symmetry) is unstable and the turbulence will self-modulate according to one of the possible soliton-like solution. In a time picture, we can say that the poloidal symmetry will be lost and the turbulence amplitude will develop a poloidal asymmetry which grows exponentially in time. The onset of the poloidal asymmetry will induce an asymmetric rate of transport on the poloidal direction [13]. When the particle diffusion is poloidally asymmetric and the local particle confinement time is shorter than the damping time of the poloidal rotation, the poloidal rotation is

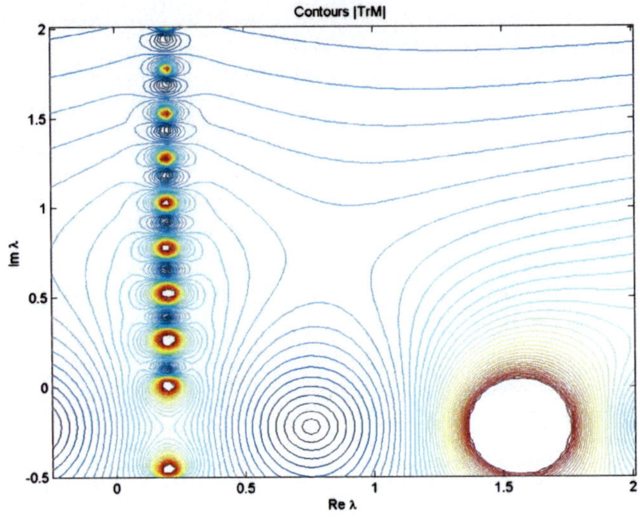

Figure 19: The discriminant $TrM(\lambda)$ of the $\mathrm{sech}(y)$ solution, to compare with the plane wave solution's discriminant.

unstable. The plasma can spontaneously spin-up. The growth rate of the asymmetry (which is given by the rate of departure from the uniform solution toward a soliton-like solution) determines the growth rate of the torque applied to plasma in the poloidal direction but the relation is not direct, depending essentially of the momentum balance in the poloidal direction. Any torque acting to generate a poloidal rotation must overcome the neoclassical magnetic pumping, which is an efficient damping mechanism. A balance of these forces depends on the comparison between the rates of (1) asymmetry growth, the radial variation of the asymmetry-induced terms, on one hand, and (2) rate of damping through magnetic pumping, on the other hand.

For the simplest case of the space-uniform envelope solution, we have the instability has the *linear* growth rate

$$i\gamma^{env} = i\mathrm{Im}\Omega = i\left|\sqrt{\alpha}k\left(\alpha k^2 - 4p^2\sigma\right)^{1/2}\right| \tag{225}$$

where $k < 2p\sqrt{\sigma/\alpha}$ is here the wavenumber of the perturbation of the envelope. In the case that has served as example in the preceeding Section, we have $p = \sqrt{t_0\omega_{unif}^{phys}/(2\sigma)} \approx 1.86$ and $k < 7.08$. Taking for example $k = 3$ we obtain $\gamma^{env} = 7.67$ or $\gamma_{phys}^{env} = 7.67 \times 10^4$ (s^{-1}).

6.2 The Soliton Developing From the Initial Perturbation

Since we know that the uniform state can be unstable we have to investigate the problem of the evolution of the system after a perturbation is applied and look for stable solutions. The solitons of this system can be realisable in real conditions since they are intrinsically more stable. We will look for soliton-like solution for the uncoupled equations, taking a single NSE like (188). The time dependence can be

$$\phi = \exp\left(i\mu^2 \alpha t\right) w\left(x\right) \tag{226}$$

giving the reduced equation

$$\frac{\partial^2 w}{\partial x^2} - \mu^2 w + \frac{\sigma}{\alpha} w^3 = 0 \tag{227}$$

where μ is a real constant. This equation has the following solution

$$w = \sqrt{\frac{2\alpha}{\sigma}} \mu \text{sech}\left(\mu x\right) \tag{228}$$

The constant μ can be related to the modulation wavenumber whose upper bound is (194). Taking $\mu \simeq 3$ and $\alpha = 0.399$ we obtain the frequency of steadily oscillating soliton $\omega_{osc}^{env} \equiv \mu^2 \alpha \simeq 1.2$, which is slower than the growth rate of the deformation, $\gamma^{lin} = 7.67$ obtained above (all in absolute units). Since the rate of change of the poloidal velocity (which will be calculated below) has comparable magnitude with ω_{osc}^{env} we conclude that the rotation has the time to develop before the oscillation reverses the sense of the applied torque. The maximum amplitude of the deformation is $\sqrt{2\alpha/\sigma}\mu = 2.224$. Large smooth deformations (small μ) with less localised profile will have even slower steady oscillatory motion and however significant amplitude, and we can expect they are favored in real situations.

6.3 The Poloidal Asymmetry of the Diffusion Fluxes

Our approach, which is adapted to the investigation of the stability of the turbulence envelope, can only provide a *model* of the turbulent field, according to Eq.(49). This is composed of the primary modes propagating in the poloidal direction with amplitude modulated by the slowly varying envelopes $A_{1,2}$ obeying the system of coupled Nonlinear Schrödinger Equations (98) and (99). In the radial direction the field has a variation given by the functions $\varphi_{1,2}\left(x\right)$ obtained by solving the equations (50). We will need to reconstitute the turbulent field in order to calculate the modified diffusion coefficient. We choose the following set of parameters:

$$\begin{aligned} a &= 1 \ (m) \\ B_T &= 3.5 \ (T) \\ T_e &= 1 \ (KeV) \ , \ T_i = 1 \ (KeV) \end{aligned} \tag{229}$$

$$\begin{aligned} n &= 10^{20} \ \left(m^{-3}\right) \\ L_n &= 0.1 \ (m) \end{aligned} \tag{230}$$

$$U_0 = 6 \times 10^5 \ (m/s) \tag{231}$$

The extension on the radial direction of the mode is taken $L_x = 0.1 \ (m)$ and this is also the normalising length. Other quantities are: normalising time $t_0 = 0.16 \times 10^{-6} \ (s)$ and $\Omega_i = 0.335 \times 10^9 \ (s^{-1})$, the ion-cyclotron frequency. From this it results the two parameters of the barotropic form of the ion mode, $\beta = 111.75$ and $\varepsilon = 0.476 \times 10^{-2}$. We take the following frequencies for the primary poloidal modes: $\omega_1 = 0.1451 \times 10^6 \ (s^{-1})$ and $\omega_2 = 0.1931 \times 10^6 \ (s^{-1})$. From the solution of the equations (50) the two eigenvalues result: $k_1 = 43.51 \ (m^{-1})$ and $k_2 = 56.81 \ (m^{-1})$. Solving the series of differential equations described in detail in Section 3 we obtain the following constants in the coupled Nonlinear Schrödinger Equations: $\alpha_1 = 0.1756 \times 10^{-2}$, $\alpha_2 = 0.2278 \times 10^{-2}$ and respectively $\sigma_1 = 0.18272$, $\sigma_2 = 0.5110$. The frequency of the uniform oscillations, Eq.(193) is taken $p = 0.0675$ giving the maximum poloidal number for the large scale instability, Eq.(194), $\overline{m} = [5.89] = 5$. The growth rate of the modulational instability is $\gamma = 4175 \ (s^{-1})$. The expression of the field becomes

$$\begin{aligned} \phi(r,\theta) &= \phi_0 \left[(1 + A_1(\theta)\operatorname{sech}(y)) \cos(k_1 y - \omega_1 t + \delta_1) \varphi_1(r) \right. \\ &\quad \left. + (1 + A_2(\theta)\operatorname{sech}(y)) \cos(k_2 y - \omega_2 t + \delta_2) \varphi_2(r) \right] \end{aligned} \tag{232}$$

where ϕ_0 is explained below. The initial phases $\delta_{1,2}$ are arbitrary and in the numerical calculations are taken as random quantities with uniform distribution on the interval $(0, 1)$.

We notice that the equations (50) fulfilled by $\varphi_{1,2}$ are homogeneous and the amplitudes of the two functions are arbitrary. We fix the amplitude ϕ_0 with the condition that the *uniform* profile of the averaged square of the potential of the fluctuating field obtains a typical value of the diffusion coefficient, $D_0 \sim \left\langle |\phi_0|^2 \right\rangle$, which is taken $D_0 \sim 1 \ (m^2/s)$. The following two figures represent the fluctuating potential, resulting from the combination of the factors described above.

A simple way to take into account the effect of the turbulence amplitude modulation on the diffusion coefficient is to consider that locally the diffusion coefficient is proportional with the quantity $\left\langle |\phi|^2 \right\rangle$, where the averages should be taken over a statistical ensemble of realizations of the fluctuating field. This requires a carefull examination when applied to numerical calculations. We have to remember that the diffusion is a statistical process where a test particle interacting with fluctuating background field performs random displacements whose mean square grows linearly in time. The random displacements have to belong to a Gaussian statistical ensemble and this requires that all kind of elementary experiences (i.e. displacements) to be represented with the corresponding probability. This is practically irrealisable in numerical calculations since the statistical ensemble of the stochastic field configurations is necessarly finite, i.e. incomplete as a Gaussian ensemble. Any average will be made on subensembles, more or less incomplete. The collection of field configurations are obtained from time *and* from space realizations of the field. The test particle should have the possibility to manifest the *linear* mean square displacement specific to the Gaussian ensembles and this rises the question on what is the minimal spatial extension on which the average should be done. We need in any case a space region larger

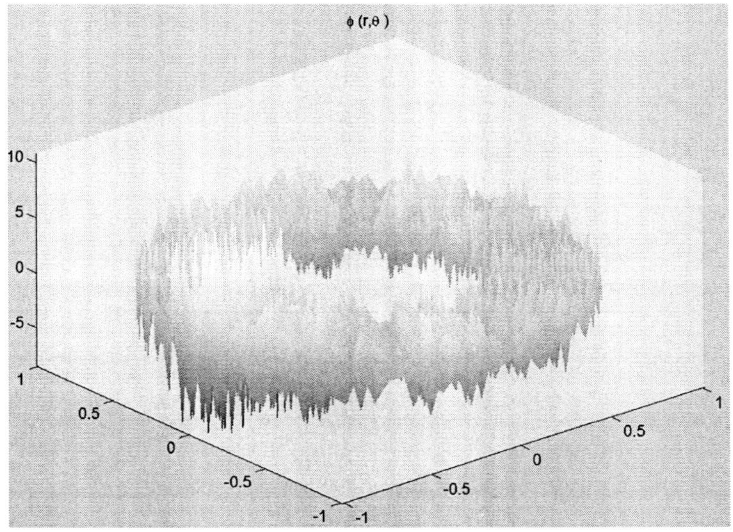

Figure 20: Instantaneous amplitude of the potential fluctuation, as calculated from the different contributions

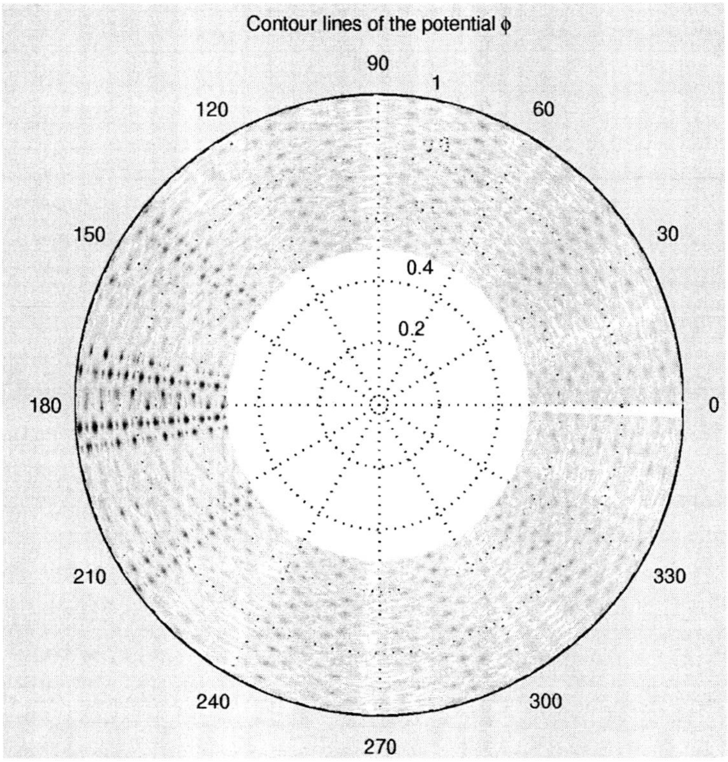

Figure 21: Contour lines of the fluctuating potential; the radial extension is artificially extended to improve the picture.

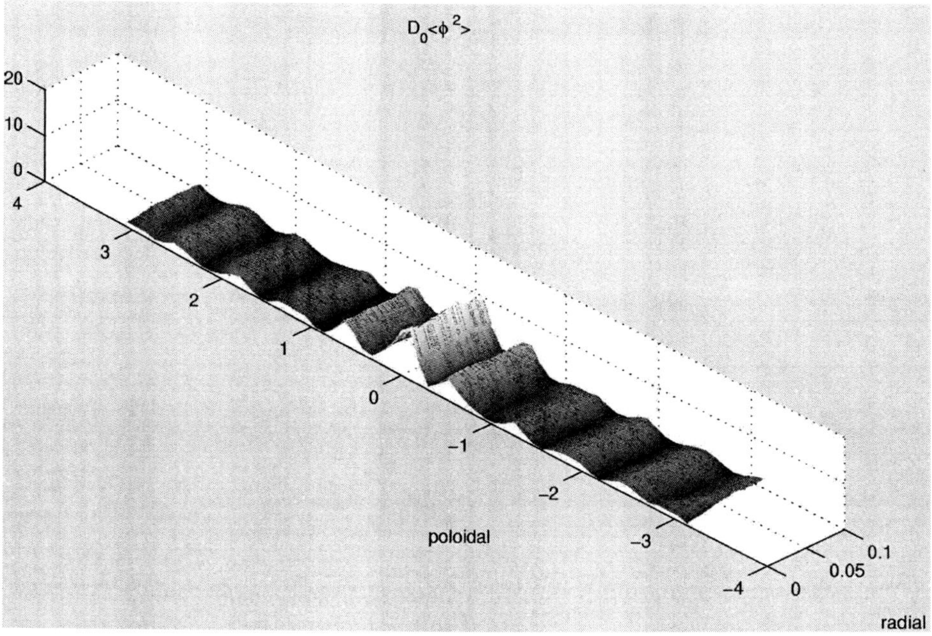

Figure 22: The radial and poloidal dependence of the average diffusion coefficient.

than the correlation length of the fluctuating field which can be taken as the inverse of the characteristic wavenumber in the field (minimum of k_1, k_2).

The profile of the quantity $D_0 \left\langle |\phi|^2 \right\rangle$ is represented in the figure below. The exemple chosed for these graphical representation is characterized by the imposed constrain to obtain a rapid variation of the functions φ_1 and φ_2 on the radial r-interval, with the intention to simulate short wavelength ion turbulence, frequently observed in the numerical simulations of the Ion Temperature Gradient driven turbulence. This is achieved by asking the numerical procedures which solves the Schrödinger-type equations (50) to look for high order κ of the eigenvalues. The pictures represented in this section correspond to $\kappa = 19$ for both eigenfunctions. This generates a potential dominated by small spatial scales and consequently, the fast variation of the local value of the diffusion coefficient. We take, in the case $\kappa = 19$ a space region containing at least three complete oscillations of the typical wavelength. This should allow at least several random steps in the typical displacement of the particle. When we do this (taking arbitrarly a factor of few units larger than the correlation length) we note that the diffusion coefficient exhibits space variations. These are natural and are inherited from the space dependence of the radial eigenfunctions $\varphi_{1,2}(r)$. Since in real situations the radial eigenfunctions are superposed with random initial phases we need to enlarge the region of averaging over most of the radial interval considered, up to a level of saturation of these spatial variation which can be considered convenient for the estimation of the poloidal torque.

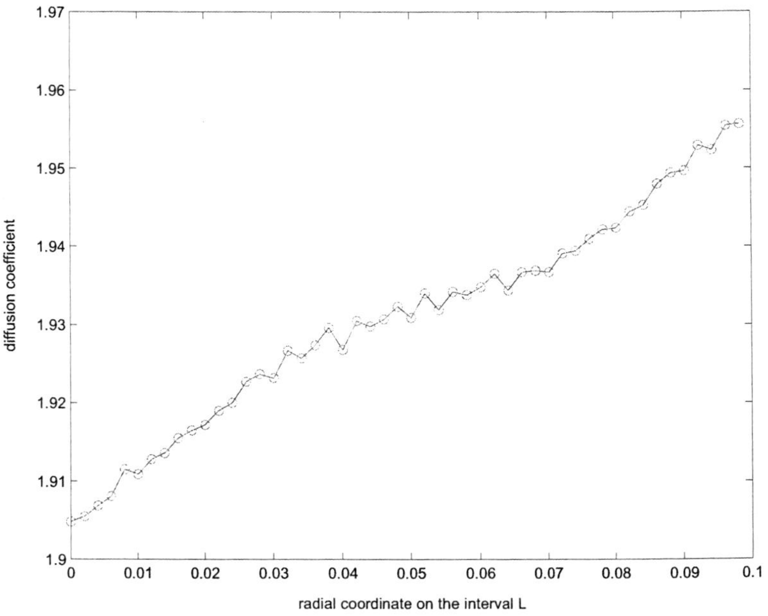

Figure 23: Radial variation of the diffusion coefficient after poloidal averaging.

6.4 The Poloidal Torque Applied on Plasma

We use the equations describing the plasma poloidal and toroidal rotation derived in [18], [19]:

$$\frac{\partial n}{\partial t}+\frac{1}{r}\frac{\partial}{\partial r}\left(rn\overline{v}_r\right)=0 \tag{233}$$

$$\frac{\partial}{\partial t}\left(nV_\varphi\right)+\frac{1}{r}\frac{\partial}{\partial r}\left[rn\left(V_\varphi\overline{v}_r-qV_\theta\widetilde{v}_r\right)\right]=0 \tag{234}$$

$$\frac{\partial}{\partial t}\left[V_\varphi+\Theta\left(1+2q^2\right)V_\theta\right]+\overline{v}_r\frac{\partial V_\varphi}{\partial r}-\widetilde{v}_r\frac{\partial}{\partial r}\left(qV_\varphi\right)+\text{magnetic pumping}=0 \tag{235}$$

The definition of the variables is $n=\langle n\left(r\right)\rangle$, $nV_\varphi\left(r\right)\equiv\left\langle nv_\varphi\frac{R}{R_0}\right\rangle$ and $V_\theta\left(r\right)=\left\langle v_\theta\frac{R}{R_0}\right\rangle$, where v_φ and v_θ are local toroidal and poloidal plasma velocities in the magnetic surface. The magnetic field is

$$\mathbf{B}=\left(0,\Theta\left(r\right),1\right)\frac{B_0\left(r\right)}{R/R_0} \tag{236}$$

with $R=R_0+r\cos\theta$ and $q=\frac{r}{R_0}\frac{1}{\Theta}$. The surface average is made according to the formula

$$\langle f\rangle=\oint\frac{d\theta}{2\pi}\frac{R}{R_0}f \tag{237}$$

In the collisional regime the *magnetic pumping* term determined by parallel viscosity is less effective and the spin up can be expected. The decay rate of the poloidal rotation by this mechanism is [20]:

$$\gamma_{MP} \simeq \frac{3}{4}\left(1+\frac{1}{2q^2}\right)^{-1}\left(\frac{l}{qR}\right)^2 \nu_{ii} \tag{238}$$

where l is the mean-free path. The velocities $\widetilde{v}_r$ and $\overline{v}_r$ arise from the diffusive flux in the radial direction, to which is associated the local radial velocity V_r. We define the two velocities

$$\overline{v}_r = \langle V_r \rangle \tag{239}$$

$$\widetilde{v}_r = \langle 2\cos\theta \; V_r \rangle \tag{240}$$

If there is no poloidal variation of the local radial velocity V_r the average in (240) is that of the trigonometric function $\cos\theta$ and the result is zero. The average will not be zero if there is a poloidal variation of the diffusion-generated particle radial velocity, $V_r(\theta)$. We can introduce the modified diffusion coefficient $\widetilde{D}$ related to the nonsymmetry,

$$n\widetilde{v}_r = \widetilde{\Gamma}_r = -\widetilde{D}\frac{\partial n}{\partial r} \tag{241}$$

with

$$\widetilde{D} = \left\langle D_0 \left\langle |\phi|^2 \right\rangle_{st} 2\cos\theta \right\rangle_{pol} \tag{242}$$

where the first averaging is statistical and the second is on the poloidal surface. The equation for the plasma poloidal velocity is

$$\Theta\left(1+2q^2\right)\left(\frac{\partial V_\theta}{\partial t} + \gamma_{MP} V_\theta\right) + qV_\theta \frac{1}{nr}\frac{\partial}{\partial r}\left(nr\widetilde{v}_r\right) = 0 \tag{243}$$

The average velocity Eq.(241) still dependes on the radial coordinate via D_0, the density $n(r)$ and its derivatives and, more important, via the space dependence of the potential $\varphi(x)$ left after the statistical averaging needed for the calculation of the diffusion coefficient (242). The following quantity must be numerically calculated

$$\gamma \equiv \frac{-q}{\Theta\left(1+2q^2\right)}\frac{1}{nr}\frac{\partial}{\partial r}\left(nr\widetilde{v}_r\right) = \frac{q}{\Theta\left(1+2q^2\right)}\frac{1}{nr}\frac{\partial}{\partial r}\left(r\left\langle D_0\left\langle |\phi|^2\right\rangle_{st} 2\cos\theta\right\rangle_{pol}\frac{\partial n}{\partial r}\right) \tag{244}$$

and this directly provides the rate of change of the poloidal velocity due to the asymmetry. This must be compared to the magnetic pumping term (238) and we expect rotation instability when

$$\gamma - \gamma_{MP} > 0 \tag{245}$$

In this formula the soliton profile of the poloidal asymmetry is considered saturated and the motion of the solitons (which would give periodic variation of the torque) is neglected. As is shown in Ref.[18] the high positive second derivative of the density profile, combined with a dependence of the diffusion coefficient D_0 on the shear of the poloidal velocity are

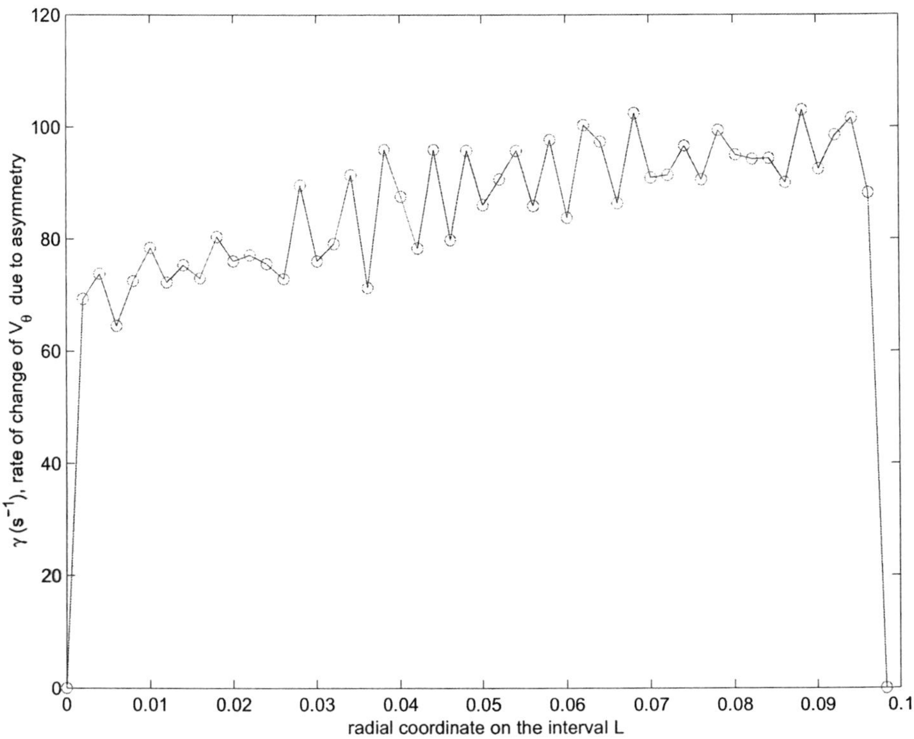

Figure 24: Radial variation of the rate of time variation γ applied on plasma in the poloidal direction, due to the asymmetry.

the favorable circumstances for the asymmetry-induced torque to overcome the magnetic pumping term and a spontaneous spin up to occur. We do not take into account the variation of D_0 with the velocity shear, but notice that the space variation of (242) also contribute to obtain a positive value for γ sufficient for being comparable to γ_{MP} and even to be larger. The most important contribution comes however from the positive-valued second density gradient, which can be seen as an indication that the spin-up can be expected in such region on the minor radius.

The values represented in the figure have been calculated using a gaussian profile for the density. The decay due to magnetic pumping is of the order of $\gamma_{MP} \sim 12 \ (s^{-1})$. We note that the torque applied to plasma shows a radial variation essentially inherited from the space variation of the statistically averaged potential fluctuations, which in general are not monotone functions of the radius on time scales of several eddy turn-over times. These spatial variations of the diffusion coefficient generate the profile of the torque which change the shear of the seed velocity and provide a characteristic which is expected for zonal flows. Even outside the domain of the parameters where the spin up is efficient compared to the magnetic pumping, the soliton asymmetry and the Stringer torque can contribute to the action of other mechanisms to generate poloidal rotation.

7 Discussion

We have examined the problem of stability of the envelope of the ion wave turbulence. Starting from a simple model for the ion dynamics and retaining the full nonlinear contribution of the ion polarization current we have derived an equation of the same form as the *barotropic* model of the atmosphere. The multiple space-time scale analysis shows that the envelope of the turbulence results from a set of coupled cubic nonlinear Schrodinger equations. The simplest solution is the space-constant envelope corresponding to a symmetric average amplitude of the turbulence on the poloidal circumference. In order to examine the stability of this solution we first review the method of inverse scattering transform in the geometric-algebraic framework, able to obtain the exact solution developing from a certain class of initial conditions (finite-band potentials). In this framework appears clear the origin of the sensitivity to the perturbations, connected to the topology of a hyperelliptic Riemann surface. The uniform solution is unstable and evolves to one of the soliton-type solution. The nonuniformity of the turbulence average profile generates a torque on the plasma, which may be sufficient to overcome the poloidal damping.

The simple physical model we have employed may be partly justified by the observation that the analysis is essentially independent of the other possible sources of poloidal asymmetry: the neoclassical variation of parameters on the magnetic surface, the toroidal effect on the nonlinearity represented as mode coupling, the effect of the particle drifts on the linear and renormalized propagators, etc. The cubic nonlinear Schrodinger equation has many solutions on the periodic domain and the relevance for the self-modulation of the envelope of the turbulence should be examined. Here we only considered the uniform envelope since its instability can be sufficient to induce plasma poloidal rotation. This rotation has an origin different of the usual sources of rotation: neoclassical, ion-loss, Reynolds stress or momentum injection. Our approach is essentially mathematical and so it operates at a more formal level. The physical origin of the instability seems to be the tendency of the system (averaged amplitude of the turbulence) to evolve to profiles characterised by the balance of the dispersion and the nonlinearity. What remains to be examined is the quantitative constraint on the minimal initial perturbation needed to produce a particular evolution: for a weak perturbation the solution may be simple "radiation" while for initially larger amplitude a soliton emerges ensuring higher robustness and a localised form. It is in this case that the asymmetry can induce plasma rotation. This question can be answered by examining the exact solution, with specific numerical tools for the determination of the main spectrum. It might be possible that the Riemann curve is not a hyperelliptic surface but a manifold close to a finite-genus curve.

References

[1] F. Spineanu and M. Vlad, *Phys. Rev. Lett.* **89,** 185001 (2002).

[2] T. E. Stringer, *Phys. Rev. Lett.* **22,** 770 (1969).

[3] S. Champeaux, P.H. Diamond, *"Streamer and zonal flow generation from envelope modulations in drift wave turbulence"*, *Phys. Lett. A* **288** (2001)214-219.

[4] K. Itoh, S.-I. Itoh, M. Yagi and A. Fukuyama, *"Solitary Radial Electric Field Structure in Tokamak Plasmas"*, *Report NIFS-525*, Dec.1997.

[5] K.H. Spatschek, S.K. Turitsyn and Yu.O. Kivshar, *"Average Envelope Soliton Dynamics in Systems with Periodically Varying Dispersion"*, *Phys. Lett. A* **204**, 269-273 (1995).

[6] E. W. Laedke, O. Kluth and K. H. Spatschek, *"Families of Discrete Solitary Waves in Nonlinear Schrödinger Equations"*, in E. Alfinito, M. Boiti, L. Martina and F. Pempinelli, *eds.* (World Scientific, Singapore, 1996) 521-528.

[7] A. V. Tur, A. V. Chechkin and V. V. Yanovsky, *Phys. Fluids B* **4**, 3513 (1992).

[8] W. Horton, *Phys. Reports* **192**, 1 (1990).

[9] B. Tan, *J. Atmospheric Sciences* **53**, 1604 (1996).

[10] B. Tan and J. P. Boyd, *Chaos, Solitons & Fractals* **11**, 1113 (2000).

[11] V. I. Pulov, I. M. Uzunov and E. J. Chacarov, *Phys. Rev. E* **57**, 3468 (1998).

[12] J. Yang, *Physica D* **108**, 92 (1997).

[13] See file: *www.ipp.cas.cz/tokamak/turbmovie.html.*

[14] E. R. Tracy and H. H. Chen, *Phys. Rev. A* **37**, 815 (1988).

[15] S. Novikov, S. V. Manakov, L. P. Pitaevskii and V. E. Zakharov, *Theory of Solitons: The Inverse Scattering Method*, Consultants Bureau, New York, 1984.

[16] M. G. Forest and D. W. McLaughlin, *J. Math. Phys.* **23**, 1248 (1982).

[17] A.R.Osborne, *PR-E* **48**, 296 (1993).

[18] A.B. Hassam and T.M. Antonsen, Jr., *"Poloidal spin-up of tokamak plamas from poloidal asymmetry of particle and momentum sources"*, *Phys. Plasmas* **1** (1994) 337.

[19] A.B. Hassam and J. F. Drake, *"Spontaneous poloidal spin-up of tokamak plasmas: Reduced equations, physical mechanisms, and sonic regimes"*, *Phys. Fluids B* **5** (1993) 4022.

[20] A.B. Hassam and R.M. Kulsrud, *"Time evolution of mass flow in a collisional tokamak"*, *Phys. Fluids* **21**, 2271 (1978).

In: Advances in Soliton Research
Editor: L.V. Chen, pp. 105-133
ISBN 1-59454-769-6

Chapter 3

QUANTUM DARK SOLITON IN BOSE-EINSTEIN CONDENSATE

Jacek Dziarmaga and Krzysztof Sacha
Intytut Fizyki Uniwersytetu Jagiellońskiego, Reymonta 4,
30-059 Kraków, Poland

Abstract

In a mean-field approximation Bose-Einstein condensates are described by the Gross-Pitaevskii equation. In quasi-one-dimensional traps this non-linear Schrödinger equation becomes one dimensional and it has dynamically stable dark soliton (kink) solutions. The classical dark soliton propagates across the condensate without any dispersion. However, the Gross-Pitaevskii equation is a mean-field approximation which neglects any depletion of atoms from the condensate wave function. We show that in the case of dark soliton atomic collisions can lead to substantial depletion from the condensate to the notch of the dark soliton wave function. In the Bogoliubov vacuum state the density of depleted atoms in the soliton notch is, for typical experimental parameters, as large as 14% of the condensate density. This minimal stationary quantum depletion indicates that depletion effects are strong. Indeed, we find that in a dynamical process of experimental generation of the dark soliton the atoms depleted from the condensate to the soliton notch fill up the notch after just 10 milliseconds. This time is in a striking coincidence with the time when the dark solitons are experimentally observed to loose contrast. This prediction is based on an expectation value of the density operator. The expectation value is an average over many experimental realizations. To see what is a generic outcome of a single realization of density measurement we derive N-particle wave functions in particle representation and then simulate density measurements using N-particle probability distributions that follow from the N-particle wave functions. We find that in the case of highly idealized states, like the Bogoliubov vacuum, a single realization of density measurement will always find an empty dark soliton with a random position that fluctuates from experiment to experiment. However, in a generic non-stationary state a single realization of the density measurement will find a dark soliton with a random position but also with a random amount of depleted atoms inside the soliton notch. Pictures of the dark soliton in a generic non-stationary state are loosing contrast.

1 Introduction

The one-dimensional Schrödinger equation for a free particle

$$i\hbar\partial_t\psi = -\frac{\hbar^2}{2m}\partial_x^2\psi. \quad (1)$$

has a nonlinear dispersion relation $E = p^2/2m$ which means that any initially localized wave packet must disperse in course of time evolution. The dispersion is prevented if an additional non-linear term is present in the Schrödinger equation

$$i\hbar\partial_t\psi = -\frac{\hbar^2}{2m}\partial_x^2\psi + g_0|\psi|^2\psi. \quad (2)$$

As is well known, equation (2) has soliton solutions both for $g_0 > 0$ and $g_0 < 0$. Here we will focus on the case of repulsive interactions $g_0 > 0$ when Eq. (2) has a kink solution which propagates without any change of shape. For vanishing soliton velocity the wave function $\psi(x)$ is stationary and it is proportional to

$$\tanh\left(\frac{x}{l_0}\right). \quad (3)$$

Solitons have been investigated experimentally in non-linear optics [1] where an effective equation for a pulse propagating in the medium is the non-linear Schrödinger equation. They appear dark on the pictures, hence their name of dark solitons.

Recently a new class of experiments became available. Developments in cooling and trapping of atoms lead to the achievement of Bose-Einstein condensation in dilute atomic gases [2]. In a perfect Bose-Einstein condensate (BEC) all atoms occupy the same condensate wave function which, in the mean field approximation, is a solution of the non-linear Schrödinger equation, known in this field as the the Gross-Pitaevskii equation [3]. This opens up a possibility of preparation and manipulation of solitons in degenerate atomic gasses and such experiments have already been carried out [4]. A soliton is prepared in two steps. In the first step a condensate is prepared in a ground state when all atoms occupy the same wave function which is the ground state solution of the Gross-Pitaevskii equation. In the second step, the so called phase imprinting procedure [5], a short and strong laser pulse is focused on the atomic cloud. In this short period of time the Hamiltonian of interacting atoms in a trap can be neglected as compared to the interaction of atoms with the strong electromagnetic wave, thus

$$i\hbar\partial_t\psi(x,t) = I(x)\psi(x,t), \quad (4)$$

where $I(x)$ is the position dependent intensity of the laser. When $I(x) = I_0\theta(x)$ (where $\theta(x)$ is the step function) and the intensity I_0 and interaction time Δt are tuned so that $I_0\Delta t/\hbar = \pi$, then the initial even wave function of the ground state, $\psi_0(x,t) = \psi_0(-x,t)$, is turned into the odd function

$$\psi(x,t+\Delta t) = e^{-i\pi\theta(x)}\psi_0(x,t), \quad (5)$$

which mimics the soliton wave function (3) (for parameters corresponding to the experiments l_0 is much less than the size of the atomic cloud).

Solitons prepared in the Hannover experiment [4] show the expected non-dispersive behavior, but only for a short period of time before their pictures loose any contrast. There is a debate in the literature about the reason for this "graying" of the dark soliton [6]. Here we will present a theory, developed in Refs.[7], which invokes quantum many body effects to explain the observed graying.

The dark soliton is a solution of the Gross-Pitaevskii equation. However, this equation is only a mean field approximation to the full problem of N interacting atoms which is really governed by an underlying N-body Schrödinger equation. The Gross-Pitaevskii equation takes into account interaction between atoms but neglects the fact that in the presence of the interaction not all atoms occupy the same single particle wave function. A fraction of atoms is always depleted from the condensate wave function. In Sec. 2 we put the dark soliton state in the context of trapped atomic gases. In Sec. 3 we employ the Bogoliubov theory to show that atoms are depleted from the condensate to the hole in the dark soliton density distribution. This indicates that the depletion may explain the observed graying. However, this conclusion is based on an average (single particle) density distribution. This average density tells us about an average over many density measurements, but a density distribution obtained in a single measurement can be qualitatively different from the average. The generic outcomes of density measurements can be generated from the N-body probability density corresponding to the N-body wave function of atoms in a trap. In Sec. 4 we calculate this N-body state and use the N-body probability distribution to generate outcomes of density measurements.

2 The Dark Soliton

Atomic gas trapped in a harmonic potential,

$$V(\vec{r}) = \frac{m}{2}(\Omega_x^2 x^2 + \Omega_y^2 y^2 + \Omega_z^2 z^2), \tag{6}$$

and interacting via two-body collisions is described, in the second quantization formalism, by the Hamiltonian

$$\hat{H} = \int d^3r\, \hat{\psi}^\dagger \left[-\frac{\hbar^2}{2m}\nabla^2 + V(\vec{r}) + \frac{g_0}{2}\hat{\psi}^\dagger\hat{\psi} \right] \hat{\psi}. \tag{7}$$

Here m is atomic mass, Ω_i are trap frequencies, and g_0 is an interaction strength given by

$$g_0 = \frac{4\pi\hbar^2 a}{m}, \tag{8}$$

where a is the s-wave scattering length of atoms.

In the limit of perfect condensate, all atoms occupy the same single particle (condensate) wave function $\psi(t,\vec{r})$ that is a solution of a time-dependent Gross-Pitaevskii equation (GPE) [3]

$$i\hbar\partial_t\psi = -\frac{\hbar^2}{2m}\nabla\psi + V(\vec{r})\psi + g_0|\psi|^2\psi. \tag{9}$$

The GPE is a classical version of the second quantized theory with a Hamiltonian (7) — substitution

$$\hat{\psi}(t,\vec{r}) \longrightarrow \psi(t,\vec{r}), \tag{10}$$

transforms the Heisenberg equation for evolution of the $\hat{\psi}$ operator to Eq. (9). When the condensate is confined in a quasi one-dimensional harmonic trap, i.e. for $\Omega_x \gg \Omega_y, \Omega_z$, a full three-dimensional condensate wave function can be factorized as

$$\psi(t,\vec{r}) = \phi(t,x)\psi_0(y,z), \tag{11}$$

where $\psi_0(y,z)$ is the ground state in the transversal direction and $\phi(t,x)$ fulfills one-dimensional GPE

$$i\hbar\partial_t\phi = -\frac{\hbar^2}{2m}\partial_x^2\phi + \frac{1}{2}m\Omega_x^2x^2\phi + g|\phi|^2\phi\,. \tag{12}$$

The effective one-dimensional interaction strength is

$$g = g_0\int dydz\,|\psi_0(y,z)|^4. \tag{13}$$

It is convenient to rewrite the GPE in the oscillator units, i.e.

$$\begin{aligned} t &\longrightarrow \frac{t}{\Omega_x} \\ x &\longrightarrow x\sqrt{\frac{\hbar}{m\Omega_x}} \\ \phi &\longrightarrow \phi\sqrt{\frac{m\Omega_x}{\hbar}} \\ g &\longrightarrow g\,\frac{N\hbar^2}{m}, \end{aligned} \tag{14}$$

where N is a number of atoms in the trap. In the new units the GPE becomes

$$i\partial_t\phi = -\frac{1}{2}\partial_x^2\phi + \frac{1}{2}x^2\phi + g|\phi|^2\phi\,. \tag{15}$$

$\phi(t,x)$ is normalized to 1, i.e. $\int_{-\infty}^{+\infty} dx\,|\phi(t,x)|^2 = 1$. In the following we will only use the oscillator units.

The "ground state" of equation (15) is a stationary state $\phi_0(x)\exp(-i\mu_0 t)$ with a symmetric $\phi_0(x)$, $\phi_0(x) = \phi_0(-x)$. In the Thomas-Fermi regime, when $g \gg 1$, ϕ_0 can be approximated by

$$\phi_0(x) \stackrel{g\gg 1}{\cong} \sqrt{\frac{2\mu_0 - x^2}{2g}}\,. \tag{16}$$

This wave function is normalized to 1 provided that $\mu_0 = (3g/2)^{2/3}/2$. It is localized and varies on a length-scale of

$$R = \sqrt{2\mu_0} = \left(\frac{3g}{2}\right)^{1/3}, \tag{17}$$

the so called Thomas-Fermi radius.

For a homogeneous condensate, i.e. in the absence of the trapping potential, the GPE possesses a solitonic solution

$$\phi(x,t) = e^{-i\rho_0 g t/\hbar}\sqrt{\rho_0}\times \left[i\frac{v}{c} + \sqrt{1-\frac{v^2}{c^2}}\tanh\left(\frac{x-q}{l_0}\sqrt{1-\frac{v^2}{c^2}}\right)\right], \tag{18}$$

where ρ_0 is a condensate density, $c = \sqrt{\rho_0 g}$ is a speed of sound and

$$l_0 = \frac{1}{\sqrt{\rho_0 g}} \tag{19}$$

is a healing length. The solution (18) describes a density minimum propagating without a change of the shape with a constant velocity, $q = vt$. For a static soliton with $v = 0$ condensate density vanishes in the notch of the soliton wave function.

In the presence of the trapping potential, but in the Thomas-Fermi regime when $g \gg 1$, the condensate density around the trap center changes very slowly on a length scale set by the large Thomas-Fermi radius (17). In first approximation the condensate can be considered uniform near the trap center and this is why the solution of the GPE can be approximated by (18) multiplied by the Thomas-Fermi profile (16). On these grounds we can expect that the antisymmetric stationary state of the Gross-Pitaevskii equation $\phi_1(x)\exp(-i\mu_1 t)$ is the dark soliton localized in the center of the trap times the Thomas-Fermi profile of the ground state (16),

$$\phi_1(x) \overset{g\gg 1}{\cong} \phi_0(x)\tanh\left(\frac{x}{l_0}\right), \tag{20}$$

where

$$l_0 = \frac{1}{\sqrt{g|\phi_0(0)|^2}} = \sqrt{2}\left(\frac{2}{3g}\right)^{1/3} \tag{21}$$

is a healing length of the condensate at the condensate peak density in the center of the trap. The soliton solution in Eq.(20) has a notch at $x = 0$. We note that with increasing g the Thomas-Fermi radius R is growing while the healing length l_0 tends to zero. For $g \gg 1$ the width of the soliton is very small as compared to the size of the condensate, $l_0 \ll R$, and our assumption of locally homogeneous condensate is justified in this regime.

Quasi one-dimensional trap is essential from the point of view of dynamical stability of the solitonic solution. Basically the soliton is stable provided excitation energies in the transversal degrees of freedom are higher than kinetic energy deposited in the soliton notch.

3 The Bogoliubov Theory

Even dynamically stable solution of the GPE is still zero order approximation of the full quantum many body problem only. The GPE takes into account interaction between atoms (in a sense of the mean field approximation) but it neglects depletion of atoms from the condensate wave function. To make one step further we can employ the Bogoliubov theory of quantum fluctuations around the mean field solution [8].

In the Bogoliubov theory [8] the bosonic field operator $\hat{\psi}$ is expanded in small quantum fluctuations $\delta\hat{\psi}$ around its nonzero expectation value

$$\hat{\psi}(x) \;=\; \langle\hat{\psi}\rangle + \delta\hat{\psi}(x) \;=\; \sqrt{N}\phi(x) + \delta\hat{\psi}(x)\,. \tag{22}$$

Here $\phi(x)$ is any stationary solution of the GPE. Substitution of Eq.(22) to the one-dimensional version of the Hamiltonian of the system

$$\hat{H} \;=\; \int dx\,\hat{\psi}^\dagger\left(-\frac{1}{2}\partial_x^2 + \frac{x^2}{2} + \frac{g}{2}\hat{\psi}^\dagger\hat{\psi} - \mu\right)\hat{\psi} \tag{23}$$

and expansion of the Hamiltonian to second order in $\delta\hat{\psi}$ gives a Hamiltonian which is quadratic in the quantum fluctuations. Note that $\langle\hat{\psi}\rangle \neq 0$ implies state of the system that is a superposition of states with different numbers of atoms. The quadratic Hamiltonian is brought to a diagonal form

$$H \;=\; \sum_m \omega_m \hat{b}_m^\dagger \hat{b}_m \;+\; \text{const} \tag{24}$$

by a Bogoliubov transformation

$$\delta\hat{\psi}(x) = \sum_m \left[\hat{b}_m u_m(x) + \hat{b}_m^\dagger v_m^*(x)\right]\,. \tag{25}$$

The operators $b_m^{(\dagger)}$ are annihilation(creation) operators of bosonic Bogoliubov quasi-particles. The mode wave functions (u_m, v_m) are solutions of the stationary Bogoliubov-de Gennes equations [3, 8]

$$\begin{aligned} -\frac{1}{2}\partial_x^2 u + \frac{1}{2}x^2 u + 2g|\phi|^2 u - \mu u + g\phi^2 v \;&=\; +\omega u\,, \\ -\frac{1}{2}\partial_x^2 v + \frac{1}{2}x^2 v + 2g|\phi|^2 v - \mu v + g\phi^{*2} u \;&=\; -\omega v\,. \end{aligned} \tag{26}$$

In order to satisfy bosonic commutation relations, $[\hat{b}_n, \hat{b}_m^\dagger] = \delta_{nm}$, the modes must be normalized so that $\langle u_n|u_m\rangle - \langle v_n|v_m\rangle = \delta_{nm}$. The Bogoliubov Hamiltonian (24) has different eigenstates that can be numbered by a number of quasi-particles in each Bogoliubov mode. The quasi-particle vacuum (Bogoliubov vacuum) is a state annihilated by all annihilation operators $\hat{b}_m$

$$\hat{b}_m|0\rangle_b = 0. \tag{27}$$

We are interested in solutions of the Bogoliubov-de Gennes equations on the soliton classical background when $\phi = \phi_1$ and $\mu = \mu_1$. There are infinite number of normalizable solutions with $\omega > 0$, the wave functions of these modes spread over the whole trap and can be identified as phonons. However, there is one solution with negative ω [9], the so called anomalous mode. In the Thomas-Fermi regime the negative ω can be approximated by [9]

$$\omega \overset{g\gg 1}{\cong} -\frac{1}{\sqrt{2}}, \tag{28}$$

i.e. $-1/\sqrt{2}$ times the frequency of the harmonic trap. This Thomas-Fermi limit of ω follows from effective equations of motion for a position of a dark soliton and from a perturbation ($g \gg 1$) treatment of the Bogoliubov equations [9]. The normalized mode corresponding to negative ω, in the Thomas-Fermi regime, reads

$$f_+(x) = u(x) + v(x) \overset{g\gg 1}{\cong} \frac{\sqrt{3g}}{2\sqrt{2}\,\cosh^2(x/l_0)} \tag{29}$$

$$f_-(x) = u(x) - v(x) \overset{g\gg 1}{\cong} \sqrt{\frac{2}{3}}\,\phi_0(x)\,. \tag{30}$$

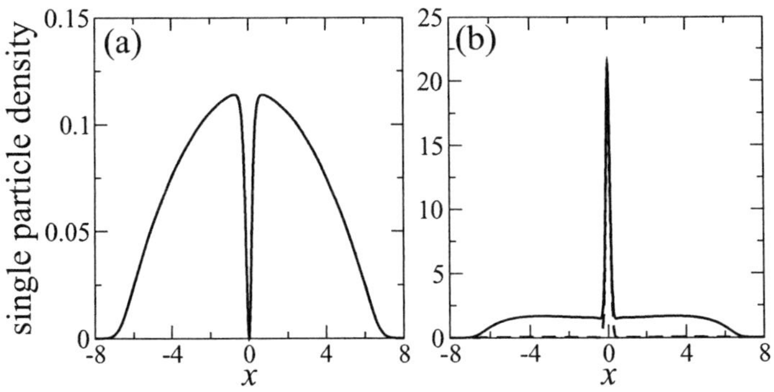

Figure 1: Numerical results for single particle density functions corresponding to $g = 200$. Panel (a) shows the density of the soliton wave function $|\phi_1(x)|^2$. In panel (b) solid line indicates the total density of incoherent atoms, (i.e. atoms depleted from the condensate wave function) while the dashed line is the density of atoms depleted to the anomalous mode.

Note that $f_+(x)$ is localized within the notch of the dark soliton (20). Such a strong localization in the notch is a property of the anomalous mode only (see Fig. 1 for numerical results of coherent and incoherent atoms density distributions in the Bogoliubov vacuum state (27) for $g = 200$). It favors the anomalous mode among other modes. At zero temperature the number of atoms depleted from the condensate wave function is very small as compared to the total number of atoms in a condensate for parameters corresponding to a typical experiment. Thus simple atomic density measurement is not able to detect quantum depletion effects unless depleted atoms are localized in the notch where condensate atoms are not present. It turns out that number of atoms depleted to the anomalous mode can be quite small but because of its strong localization the effect is quite dramatic. We will argue that depletion of atoms from the condensate fills up the notch in the condensate wave function (20) with incoherent atoms. This effect may appear experimentally as graying of the dark soliton. In the following we truncate the Bogoliubov spectrum to the anomalous mode alone because from all the modes the anomalous mode will contribute the most to the density of incoherent atoms filling the notch.

Before we proceed further, let us show that (29-30) are indeed approximate solutions of the Bogoliubov-de Gennes equations with $\omega \simeq -1/\sqrt{2}$. Eqs.(26) can be combined into

$$-\frac{1}{2}\partial_x^2 f_+ + \frac{1}{2}x^2 f_+ + 3g\phi_1^2 f_+ - \mu_1 f_+ = \omega f_- \,, \tag{31}$$

$$-\frac{1}{2}\partial_x^2 f_- + \frac{1}{2}x^2 f_- + g\phi_1^2 f_- - \mu_1 f_- = \omega f_+ \,. \tag{32}$$

The anomalous mode is a perturbation approximation to the motion of the dark soliton with respect to the background condensate. In the limit of $g \to \infty$ the Thomas-Fermi radius of the condensate (17) is growing to infinity and at the same time the healing length (21) is shrinking to 0. In this limit the condensate is uniform on the length-scale of the soliton width and the anomalous mode of Eqs.(31,32) becomes

$$\begin{aligned} \lim_{g\to\infty} f_+(x) &\propto \frac{1}{\cosh^2(x/l_0)} \,, \\ \lim_{g\to\infty} f_-(x) &= 0 \,. \end{aligned} \tag{33}$$

This $f_+(x)$ is simply a translational mode of the soliton in an almost uniform condensate. This limit gives us the leading $g \gg 1$ term in the expression for $f_+(x)$.

We use Eq.(32) to find the leading term in $f_-(x)$. To begin with we note that $\phi_1(x)$ satisfies

$$-\frac{1}{2}\partial_x^2 \phi_0 + \frac{1}{2}x^2 \phi_0 + g\phi_0^3 - \mu_0\phi_0 = 0 \,. \tag{34}$$

Comparing Eq.(34) with Eq.(32) one may conclude that $f_-(x) \cong \alpha\phi_0(x)$ with a certain constant α. Substitution of this ansatz and of leading $f_+(x)$ from Eqs.(33) into Eq.(32) and then using Eq.(34) gives an equation

$$\begin{aligned} &-(\mu_1 - \mu_0)\alpha\phi_0(x) \, + \\ &g\left[\phi_1^2(x) - \phi_0^2(x)\right]\alpha\phi_0(x) = \frac{\omega}{\cosh^2(x/l_0)} \,. \end{aligned}$$

We expect that with the right choice of α this equation is approximately satisfied for $g \gg 1$. As $\mu_1 - \mu_0 = \mathcal{O}(g^0)$ the first term on the left hand side is small as compared to the second term and can be neglected. The second term is localized on a healing length around $x = 0$ and the slow factor $\phi_0(x)$ in this term can be replaced by $\phi_0(0)$. Finally we can use the $g \gg 1$ formulas in Eqs.(16,20) to get

$$g\left[\tanh^2(x/l_0) - 1\right]\alpha\phi_0^3(0) = \frac{\omega}{\cosh^2(x/l_0)} . \tag{35}$$

This equation is satisfied by

$$\alpha = -\frac{\omega}{g\phi_0^3(0)} = \frac{4}{3\sqrt{g}} , \tag{36}$$

which defines the leading term in $f_-(x)$. After normalization, so that $\int dx\ (|u|^2 - |v|^2) \equiv \int dx\ f_+ f_- = 1$, we get (29-30).

Having constructed $f_\pm = u \pm v$ we can calculate an operator corresponding to a number of depleted atoms in the anomalous mode

$$d\hat{N}(t) = \int dx\ \delta\hat{\psi}^\dagger(x,t)\ \delta\hat{\psi}(x,t) , \tag{37}$$

where

$$\delta\hat{\psi}(x,t) = u(x)\ e^{-i\omega t}\ \hat{b} + v^*(x)\ e^{+i\omega t}\ \hat{b}^\dagger \tag{38}$$

is an annihilation operator of an incoherent atom in the anomalous mode. We use Heisenberg picture defined by the Bogoliubov Hamiltonian of the anomalous mode

$$H_B = -\frac{1}{\sqrt{2}}\ \hat{b}^\dagger\hat{b} . \tag{39}$$

The operator of the number of depleted atoms is

$$\begin{aligned} d\hat{N}(t) = \int dx\ [&|u(x)|^2\hat{b}^\dagger\hat{b} + |v(x)|^2\hat{b}\hat{b}^\dagger + \\ &u(x)v(x)e^{-2i\omega t}\hat{b}\hat{b} + u^*(x)v^*(x)e^{+2i\omega t}\hat{b}^\dagger\hat{b}^\dagger] . \end{aligned} \tag{40}$$

The density of incoherent atoms is the expectation value of the integrand.

For a condensate in a ground state like in Eq.(16) the Bogoliubov spectrum has only positive frequencies (i.e. there are no anomalous modes). The state with no quasi-particles (27), the Bogoliubov vacuum, is the ground state of the Bogoliubov Hamiltonian and it is also the $T = 0$ state of thermodynamic equilibrium [10]. If the solitonic state (20) is the condensate wave function, the Bogoliubov vacuum does not correspond to the thermodynamic equilibrium. The vacuum is only one out of many eigenstates of the system. The uniqueness of the Bogoliubov vacuum hinges on that it is the minimal depletion state among eigenstates. That is a number of depleted atoms for an eigenstate with m quasi-particles,

$$dN = {}_b\langle m|d\hat{N}(t)|m\rangle_b = \int dx\ [|u(x)|^2 m + |v(x)|^2(m+1)], \tag{41}$$

where

$$|m\rangle_b = \frac{\left(\hat{b}^\dagger\right)^m}{\sqrt{m!}}|0\rangle_b, \tag{42}$$

is minimal when $m = 0$.

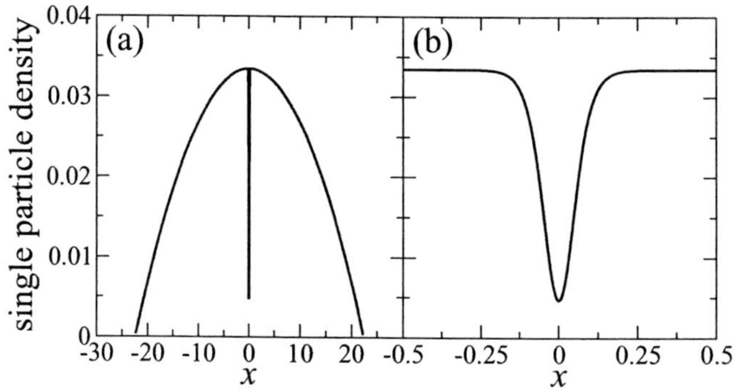

Figure 2: Single particle density function for a condensate in the Bogoliubov vacuum state for the experimental parameters [4] (the parameters correspond to $g = 7500$). Panel (a) shows the total (i.e. the sum of the coherent and incoherent parts) density function and panel (b) is a zoom on the $x \simeq 0$ part of panel (a) — note that the density of incoherent atoms at the soliton notch is about 14% of the maximal density of the surrounding condensate.

We have calculated analytically for $g \gg 1$ a number of depleted atoms in the vacuum state of the anomalous mode (the number is of course stationary because the vacuum state is an eigenstate)

$$dN_b \;=\; {}_b\langle 0|d\hat{N}(t)|0\rangle_b \;=\; \int dx\ |v(x)|^2\ , \tag{43}$$

that turns out to scale with g as

$$dN_b \propto g^{2/3}. \tag{44}$$

When we compare the density of depleted atoms in the soliton notch

$$\frac{dN_b}{l_0} \;\propto\; g \;\propto\; N\ , \tag{45}$$

(where N is the total number of atoms in the system) to the density of atoms in the condensate

$$\frac{N}{R} \propto \frac{N}{g^{1/3}} \;\propto\; N^{2/3}\ . \tag{46}$$

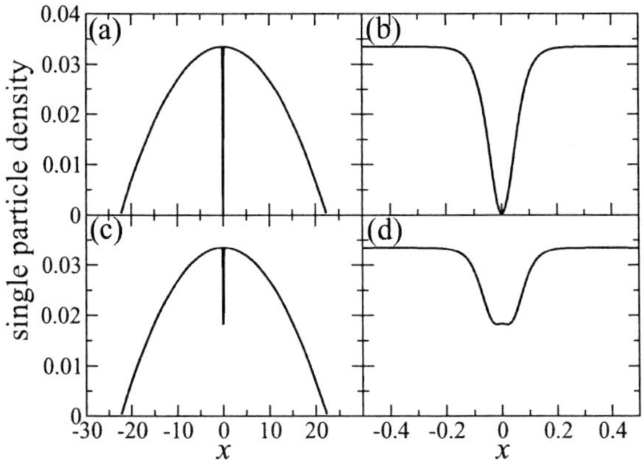

Figure 3: Panel (a): the initial ($t = 0$) single particle density function for the system initially in the minimal depletion state $|0\rangle_c$ (the values of the parameters correspond to the Hannover experiment [4]). Panel (b) is a zoom on the $x \simeq 0$ part of panel (a) — the density of the incoherent atoms at the soliton notch is about 1% of the maximal condensate density. Panels (c)-(d): the same as in the previous panels but for $t = 0.3\ (\pi/\sqrt{2}\Omega_x) = 7.6$ ms. For $t > 11$ms the density of incoherent atoms in the soliton notch becomes greater than the density of the surrounding condensate and the linear Bogoliubov theory breaks down.

we see that, for a sufficiently large total number of atoms N the density of incoherent atoms in the notch will exceed the density of the background condensate. The notch will not be visible. When the density of depleted atoms becomes comparable to the density of the condensate one we are not allowed to use the linear Bogoliubov theory any more because we cannot neglect interactions between quasi-particles.

In Fig. 2 we show total density plots in the Bogoliubov vacuum for the parameters of the Hannover experiment [4] ($N = 1.5 \cdot 10^5$ ^{87}Rb atoms in the harmonic trap with $\Omega_x = 2\pi \cdot 14$ s^{-1} and $\Omega_\perp = 2\pi \cdot 425$ s^{-1}; $g = 7500$ [11]). As we can see, even for very reasonable trap parameters and moderate N the soliton notch is substantially filled with incoherent atoms. For a larger N, the soliton would not be observed at all.

So far we have considered eigenstates of the Bogoliubov Hamiltonian, especially we have analyzed the Bogoliubov vacuum of the anomalous mode. However, the Bogoliubov vacuum is not relevant for dark solitons in current experiments. Solitons are excited experimentally employing a phase imprinting method [5, 4] on a condensate in the ground state. The phase imprinting method [5] does not prepare the soliton in the Bogoliubov vacuum state or any other eigenstate. We may assume that before the phase imprinting the system of N atoms is in its ground state. The ground state is a condensate (16) with all Bogoliubov modes in a vacuum state. There are no anomalous modes and the density of incoherent atoms is very small as compared to the density of the condensate. The ground state is very close to a perfect condensate with *all* atoms in the condensate wave function (16). We may thus approximate the initial condensate by a perfect condensate without any depletion.

For a perfect initial condensate the phase imprinting can be thought of as an instantaneous operation on the condensate wave function. Right after this operation one gets a condensate with an antisymmetric wave function $\phi(x) = -\phi(-x)$. To focus on the quantum depletion and not on classical evolution of $\phi(x)$ we assume that $\phi(x) = \phi_1(x)$ — the initial dark soliton is a (nearly) perfect condensate with the wave function (20). In the Hannover experiment [4] they create moving gray solitons while here, for the sake of simplicity, we assume a stationary dark soliton. We think that in spite of this, qualitative comparison with the experiment is still possible.

The initial state without depletion [with all atoms in solitonic wave function $\phi_1(x)$] is far from the Bogoliubov vacuum $|0\rangle_b$ or from any other eigenstates. The initial state is not stationary, its depletion may grow from zero to values which far exceed the depletion in the Bogoliubov vacuum. In the following we quantify this effect.

In the framework of the Bogoliubov theory truncated to the anomalous mode (39), the initial depletion-free soliton can be very well approximated by a variational state $|\psi\rangle$ that minimizes the initial number of atoms depleted from the background condensate $\phi_1(x)$

$$dN(0) \;=\; \langle\psi|d\hat{N}(0)|\psi\rangle \; . \tag{47}$$

To find this optimal state it is convenient to introduce bosonic operators $\hat{c}$ and $\hat{c}^\dagger$,

$$\hat{c} \;=\; \frac{1}{2}\sqrt{\frac{\varpi}{|\omega|}}\left(\hat{b}+\hat{b}^\dagger\right) + \frac{1}{2}\sqrt{\frac{|\omega|}{\varpi}}\left(\hat{b}-\hat{b}^\dagger\right) \; , \tag{48}$$

where

$$\varpi \;=\; 2^{-13/12}(3g)^{1/3} \; . \tag{49}$$

In terms of these new bosonic operators the initial number of depleted atoms (40) becomes

$$d\hat{N}(0) \;=\; \frac{2\sqrt{2}\varpi}{3}\,\hat{c}^\dagger\hat{c} \;+\; \left(\frac{\sqrt{2}\varpi}{3}-\frac{1}{2}\right) \; . \tag{50}$$

This operator is a sum of a nonnegative operator $\sim\, \hat{c}^\dagger\hat{c}$ and a constant. $\langle\psi|\hat{c}^\dagger\hat{c}|\psi\rangle$ is minimized by the unique vacuum $|\psi\rangle \;=\; |0\rangle_c$ of the bosonic annihilation operator $\hat{c}$, $\hat{c}|0\rangle_c = 0$. The minimal depletion state is $|\psi\rangle \;=\; |0\rangle_c$. This vacuum is different from the Bogoliubov vacuum $|0\rangle_b$,

$$|0\rangle_c \;\neq\; |0\rangle_b \; . \tag{51}$$

In this optimal state $|0\rangle_c$ the initial number of depleted atoms in the soliton notch scales with the number of atoms as $N^{1/3}$. The initial density of incoherent atoms in the notch

$$\frac{dN_c(0)}{l_0} \propto\; N^{2/3} \; , \tag{52}$$

here $dN_c(0) = {}_c\langle 0|d\hat{N}(0)|0\rangle_c$, grows with N in the same way as the condensate density in Eq. (46).

In Fig. 3(a,b) we show that for a realistic value of the scattering length and other parameters the minimal density of incoherent atoms in the notch is around 1% of the peak density of the surrounding condensate. Our minimal depletion state is a very accurate approximation to the initial perfect condensate with a dark soliton. As both coherent and incoherent densities grow with N in the same way, compare Eqs.(46,52), the quality of this approximation does not change with increasing N. The negligible depletion in the minimal depletion state justifies *a posteriori* our truncation of the theory to the anomalous mode.

The initial density of incoherent atoms in the notch (calculated in the $|0\rangle_c$ state) is negligible, see Fig. 3. The initial dark soliton must be very dark. This small depletion is not stationary because the initial minimal depletion state $|0\rangle_c$ is not an eigenstate of the Bogoliubov Hamiltonian (39). From Eqs. (40,48) one can easily deduce time evolution of the number of depleted atoms

$$\begin{aligned} dN_c(t) &= {}_c\langle 0|d\hat{N}(t)|0\rangle_c = \\ & dN_c(0) + \frac{(\varpi^2 - \omega^2)^2}{3\sqrt{2}\omega^2\varpi} \sin^2(\omega t) \ . \end{aligned} \tag{53}$$

The coefficient in front of the $\sin^2(\omega t)$ function scales with the total number of atoms as N which implies that the maximal density of depleted atoms in the soliton notch

$$\frac{dN_b\left(t = \pi/2|\omega|\right)}{l_0} \propto N^{4/3} \ . \tag{54}$$

Hence, for large enough N the density of depleted atoms in the soliton notch will reach the density of the condensate (46) before it reaches the maximum (54) at $t = \pi/2|\omega|$.

Eq.(53) might suggest that the system will go through a series of periodic revivals of the initial state $|0\rangle_c$ with a period of $\pi/|\omega|$. This is not a correct conclusion for large N. After a graying time

$$t_G < \frac{\pi}{2|\omega|} \ , \tag{55}$$

when the density of incoherent atoms in the notch and the density of the condensate become comparable for the first time, we cannot use the noninteracting Bogoliubov theory any more. After t_G one can ignore neither anharmonicity in the Hamiltonian (39) of the anomalous mode nor interaction of the anomalous mode with other Bogoliubov modes (phonons). We expect that these nonlinearities lead to fast dephasing and prevent any revivals. The number of depleted atoms in the notch initially follows Eq.(53) until at around the graying time t_G their density saturates at a density comparable to the density of the condensate. This will appear as graying of the dark soliton. The nonlinear stage deserves more quantitative analysis. However, we expect that, because of the global instability that shows in the negative frequency of the anomalous mode, the nonlinearities will not suppress depletion.

Eq.(55) gives a crude upper estimate of t_G, $t_G < \pi/2|\omega|$. Given Eq.(28) we obtain in real time units

$$t_G < \frac{\pi}{\sqrt{2}\Omega_x} \,. \tag{56}$$

Here Ω_x is an axial trap frequency. In the Hannover experiment of Ref.[4], $\pi/\sqrt{2}\Omega_x =$ 25 ms while the density of the incoherent atoms at the soliton notch becomes [according to Eq. (53)] comparable to the condensate density after $t_G \simeq 10$ ms, see Fig. 3. This number is close to the observed graying time of dark solitons in the experiment [4].

However, these estimates are based on the single particle expectation values of the number of depleted atoms. Single particle expectation values can be very inaccurate predictions of experiments on a Bose-condensed system if the system is not a perfect condensate because they ignore any correlation effects.

4 Density Measurement on a Bose-Condensed System

Single particle probability density gives prediction for atomic density averaged over many experiments [12]. That is if one prepares a series of identical experiments and in each experiment the position of one atom is measured then collecting all results together the density of atomic cloud will correspond to the single particle probability density. However, in a many body system atomic density can be measured in a single experiment. To predict the results of a destructive atomic density measurement we may not use the single particle density probability. The only exception is the case of a product N-particle state where all particles are described by the same single-particle wave function. Not surprisingly the particles must be bosons forming a Bose-Einstein condensate, then N-particle probability density factorizes to a product of N independent densities. If a system is not a perfect condensate to obtain predictions for a single atomic density measurement we have to choose randomly position of N atoms according to the N-body probability density [12].

To predict the measurement it is necessary to have at hand N particle state of the system. Even if we suppose this state is available we still face a challenging problem how to sample N-dimensional probability density when N is of the order of 10^5. Javanainen and Yoo [12] overcome this problem by realizing that it is much easier to choose positions of atoms one by one than to sample N-dimensional probability density. The problem considered by Javanainen and Yoo is also an excellent visualization of the fact that if a many body system is far from a condensate a density measurement can give results very distinct from the single particle density. In Ref. [12] the authors analyze interference of two perfect condensates. The two condensates are perfect but the entire system is very far from a perfect condensate, that is, the system is in a Fock state

$$|\psi\rangle = \left|\frac{N}{2}, \frac{N}{2}\right\rangle \tag{57}$$

with $N/2$ atoms in a condensate of wave function $\phi_a(x)$ and $N/2$ in a condensate with an orthogonal wave function $\phi_b(x)$. A single particle density matrix of the state (57) predicts that on average a density measurement will give a distribution of atoms

$$\rho(x) = \langle\psi|\hat{\psi}^\dagger(x)\hat{\psi}(x)|\psi\rangle = \frac{N}{2}\left(|\phi_a(x)|^2 + |\phi_b(x)|^2\right) \tag{58}$$

without any interference fringes between the two condensates. This single particle distribution is an average over many density measurements on the same Fock state (57). To predict positions of N atoms corresponding to a single experiment Javanainen and Yoo apply the following procedure. The first atom can be easily chosen according to the single particle probability density (58). Having the position of the first atom x_1 one has to annihilate this atom from the initial state

$$|\psi\rangle_1 \sim \hat{\psi}(x_1)|\psi\rangle. \tag{59}$$

Then to determine the position of the second atom one again has to calculate the single particle probability density but this time the density corresponding to the $|\psi\rangle_1$ state

$$\rho(x_2) = {}_1\langle\psi|\hat{\psi}^\dagger(x_2)\hat{\psi}(x_2)|\psi\rangle_1, \tag{60}$$

choose x_2 according to $\rho(x_2)$ and the whole procedure is repeated up to the last atom. To summarize, having an input N-particle state $|\psi\rangle$ (in a particle representation build on the two single particle modes ϕ_a, ϕ_b) and applying the described algorithm one generates a random string of coordinates $(x_1, \ldots, x_N)$ according to the N-particle probability distribution. Each application of the algorithm returns a random outcome $(x_1, \ldots, x_N)$ of a single realization of density measurement on the state $|\psi\rangle$.

In Fig. 4 there is an example of a result of a single density measurement that one can obtain in the Javanainen and Yoo model. They have chosen plane waves in a unit box as wave functions ϕ_a and ϕ_b, i.e.

$$\begin{aligned} \phi_a(x) &= e^{i\pi x}, \\ \phi_b(x) &= e^{-i\pi x}. \end{aligned} \tag{61}$$

Then the single particle density (58) turns out to be a uniform distribution without any interference fringes. However, single measurements reveal distributions with an interference term which can be parametrized by a single parameter φ

$$\begin{aligned} \frac{N}{2} \quad (\quad & |\phi_a(x)|^2 + |\phi_b(x)|^2 + \\ & + \phi_a(x)\phi_b^*(x)e^{i\varphi} + \text{c.c.}) \; . \end{aligned} \tag{62}$$

For a given experimental realization the phase φ is chosen randomly from the interval $[0, 2\pi)$. The outcome of a density measurement is the same as if before the measurement all N atoms were not in the state (57) but in an single condensate with a wave function

$$\frac{1}{\sqrt{2}}\left(\phi_a + \phi_b e^{i\varphi}\right) \tag{63}$$

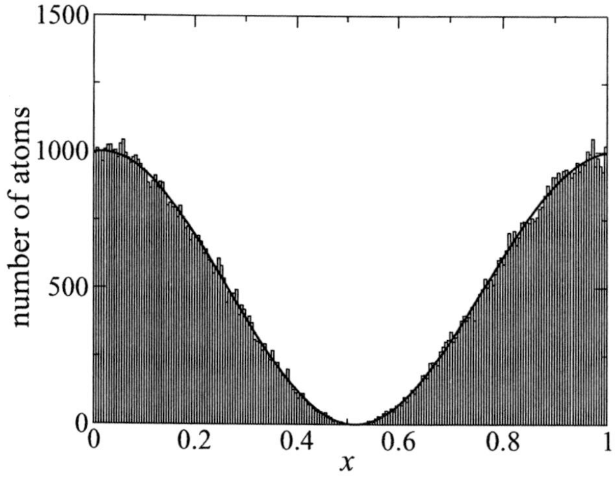

Figure 4: Density measurement for state (57) where $N = 10^5$ and the occupied modes are plane waves in a unit box, Eqs. (61). The histogram represents number of particles in bins measured in a single experiment. The solid line is the density (62) where $\varphi = -0.092$. Note that the single particle density function (58) predicts uniform distribution of atoms.

with a relative phase φ that is not known before we actually measure the density distribution. Looking at the initial state (57) such an interpretation is obviously incorrect. However, it turns out [13] that it is enough to measure only a small fraction of the large total N to prepare the remaining atoms in one of the condensate wave functions (63), or, in other words, to establish a definite phase φ. Destructive measurement of atomic positions, which effectively annihilates measured atoms from the trap, drives the remaining atoms into a condensate state with a randomly picked condensate wave function.

Let us come back now to the soliton problem. Analysis of a condensate in a solitonic state carried out in the previous section indicates that due to the interactions between atoms the condensate is not perfect — there are atoms depleted from the solitonic wave function. Moreover, the depleted atoms occupy a mode that is strongly localized in the soliton notch, i.e. in the place where there are no coherent atoms. Single particle density (corresponding to, for example, the Bogoliubov vacuum) shows substantial filling of the soliton notch by incoherent atoms. However, as we will see, from the point of view of a single measurement, such prediction is misleading. To predict possible outcomes of individual measurements on an N-particle system it is essential to know its N-particle wave function. To this end we will employ a number-conserving Bogoliubov theory [10, 14, 7]. In the previous section we have explored the traditional symmetry-breaking version of the Bogoliubov theory. While the results for a single particle density function in both the number-conserving and symmetry-breaking Bogoliubov theories are the same, in the latter case we do not obtain the N body state of the system. The key idea of the original Bogoliubov theory [8] is the $U(1)$ symmetry breaking approach where the atomic field operator is assumed to have a nonzero expectation value. This *coherent* state necessarily involves superposition of states with different numbers of atoms, an assumption very far from experimental reality. General formalism how to calculate eigenstates of the system in particle representation within the number-conserving Bogoliubov theory is shortly described in the Appendix. Here, we will use only the final results for the case of a condensate in a solitonic state.

A perfect condensate where all N atoms occupy the solitonic wave function ϕ_1 [see (20)] is

$$\left(\hat{a}_1^\dagger\right)^N |0\rangle \, , \tag{64}$$

where $\hat{a}_1^\dagger$ creates an atom in the mode ϕ_1. The wave function ϕ_1 is a solution of the Gross-Pitaevskii equation which takes into account interactions between atoms but neglects depletion of atoms from the condensate wave function. Including small quantum depletion in the framework of the number-conserving Bogoliubov theory [10] leads to pair-correlated eigenstates [14, 7]. For example, the state without any quasi-particles (the Bogoliubov vacuum) $|0\rangle_b$ has a pair-correlated form

$$|0\rangle_b \sim (\hat{d}^\dagger)^{N/2}|0\rangle = \left(\hat{a}_1^\dagger\hat{a}_1^\dagger + \sum_{s\neq 1}\lambda_s\hat{a}_s^\dagger\hat{a}_s^\dagger\right)^{N/2}|0\rangle \, . \tag{65}$$

Operators $\hat{a}_s$ annihilate in a basis of functions ϕ_s orthogonal to the condensate wave function ϕ_1. The N-particle state (65) is a foundation on which one can build theory of BEC entirely in the language of N-particle wave functions.

Solution of the Bogoliubov-de Gennes equations in the number-conserving Bogoliubov theory (see Appendix) reveals many different modes of the quasi-particle spectrum. However, in the following (similarly as in the previous section) we truncate the Bogoliubov spectrum to the anomalous mode alone because only this mode has a wave function localized in the soliton notch. The anomalous mode solution is

$$u_1 = \frac{1}{2}\left(f_+ + f_-\right) \, , \quad v_1 = \frac{1}{2}\left(f_+ - f_-\right) \, , \tag{66}$$

with functions $f_\pm$ in Eqs.(29-30). For $g \gg 1$ equations (66) are dominated by f_+, $\langle f_+|f_+\rangle \gg \langle f_-|f_-\rangle$, and the two functions become the same, $u_1 \simeq v_1 \simeq \frac{1}{2}f_+$. As a result in the Thomas-Fermi limit the anomalous mode does not mix with other modes and the dominant eigenfunction is

$$\phi_+(x) = \frac{f_+(x)}{\sqrt{\langle f_+|f_+\rangle}} \, . \tag{67}$$

This can be verified numerically for a moderate values of g. In Fig. 5 there are a few modes ϕ_s [see Eq. (65)] (found numerically for $g = 200$) that correspond to the greatest values of $|\lambda_s|$. One can see that even for not too large g the wave function ϕ_+ is a quite good approximation for the exact mode localized in the soliton notch.

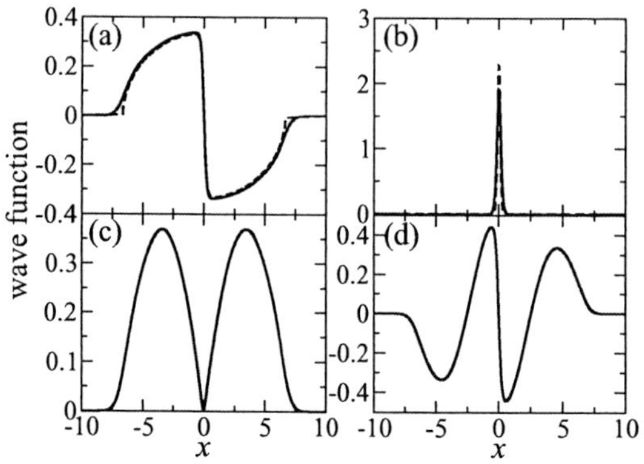

Figure 5: The condensate (solitonic) mode ϕ_1 [panel (a)] and a few modes ϕ_s [panels (b)-(d)], see (65), found numerically for $g = 200$. The presented modes are the most dominant ones, i.e. the corresponding values of $|\lambda_s|$ in (65) are the greatest — they are equaled to 0.92 [panel (b)], -0.92 [panel (c)] and -0.85 [panel (d)]. The dashed line in panel (a) is a soliton wave function in the Thomas-Fermi approximation (20) while the dashed line in panel (b) refers to ϕ_+, Eq. (67). Note that the mode localized in the soliton notch is quite well approximated by ϕ_+ despite the fact that the value of g is not very big.

The truncation of the single particle Hilbert space to the ϕ_1 and ϕ_+ modes allows us to write the Bogoliubov vacuum state in the form

$$|0\rangle_b \sim \left(\hat{a}_1^\dagger \hat{a}_1^\dagger + \lambda_+ \hat{a}_+^\dagger \hat{a}_+^\dagger\right)^{N/2} |0\rangle \tag{68}$$

To find the asymptotic behavior of λ_+ for $g \gg 1$ we have still to keep f_-. The eigenvalue can be obtained from the general prescription presented in the Appendix and it is

$$\lambda_+ \simeq \frac{\langle f_+|f_+\rangle - \langle f_+|f_-\rangle}{\langle f_+|f_+\rangle + \langle f_+|f_-\rangle} \simeq 1 - \frac{2^{7/6}3^{1/3}}{g^{2/3}} . \tag{69}$$

Knowing λ_+ we can find an average number of depleted atoms dN_b in the Bogoliubov vacuum state (68). In the particle Fock representation the unnormalized vacuum state (68) is a sum

$$|0\rangle_b \quad \sim \quad \sum_{k=0}^{\frac{N}{2}} \lambda_+^k \frac{\sqrt{(N-2k)!\,(2k)!}}{(N/2-k)!\,k!} |N-2k, 2k\rangle. \tag{70}$$

For depletion greater than unity but small with respect to the total number of atoms N, i.e. $1 \ll dN_b \ll N$, we may use the Stirling formula, and then replace the sum by an integral. As a result we find expressions for dN_b and its dispersion $D(dN_b)$

$$dN_b \simeq -\frac{1}{\ln\lambda} \simeq \frac{g^{2/3}}{4\,2^{1/6}\,3^{1/3}}\,, \tag{71}$$

$$D(dN_b) = dN_b. \tag{72}$$

It is remarkable that the dispersion of dN_b is the same as the dN_b itself. This indicates that the single particle average dN_b alone is a very crude description of the quantum state. Note also that dN_b scales with g in the same way as the corresponding quantity in the original Bogoliubov theory, compare (44).

The single particle density distribution in the vacuum state (68,70) is

$$\rho(x) = (N - dN_b)|\phi_1(x)|^2 + dN_b|\phi_+(x)|^2. \tag{73}$$

The first term on the right hand side is the dark soliton density profile with a hole at $x = 0$, while the second one stands for the depleted atoms localized in the soliton notch. Equation (73) predicts an average over many experimental realizations. For a single experiment one has to choose randomly positions of N atoms according to the N-particle probability distribution. From the results obtained by Javanainen and Yoo for the case of the interference of two Bose-Einstein condensates, we may expect that, in general, measurement a small fraction of the total number of atoms will drive the remaining atoms to a condensate wave function

$$\phi_1(x) - \phi_+(x)qe^{i\varphi}\,, \tag{74}$$

where real parameters (q, φ) are random and fluctuate from one experiment to another. Thus a single measurement of a depleted dark soliton will show the same atomic density distribution as if all N atoms were initially prepared in a condensate with one of wave functions (74). In the special case of $\varphi = 0, \pi$ the condensate wave function (74) is simply a displaced dark soliton

$$\phi_1(x) - \phi_+(x)q \simeq \phi_0(x)\tanh\left(\frac{x - q\sqrt{3l_0/4}}{l_0}\right)\,, \tag{75}$$

where ϕ_0 is the Thomas-Fermi profile, see Eq. (16). In general, when $\varphi \neq 0, \pi$, the condensate wave function (74) gives atomic density distribution

$$N|\phi_1 - \phi_+qe^{i\varphi}|^2 =$$
$$N\left[\phi_1 - \phi_+q\cos\varphi\right]^2 + N\phi_+^2q^2\sin^2\varphi\,. \tag{76}$$

For small q this distribution can be interpreted as a soliton shifted to $q\sqrt{3l_0/4}\cos\varphi$, but with a nonzero atomic density $N\phi_+^2q^2\sin^2\varphi$ in the notch. For $\varphi \neq 0, \pi$ the soliton appears as a displaced gray soliton.

Adopting the scheme of Javanainen and Yoo we generated results of individual density measurements of the Bogoliubov vacuum state (68,70). A generic outcome is a displaced

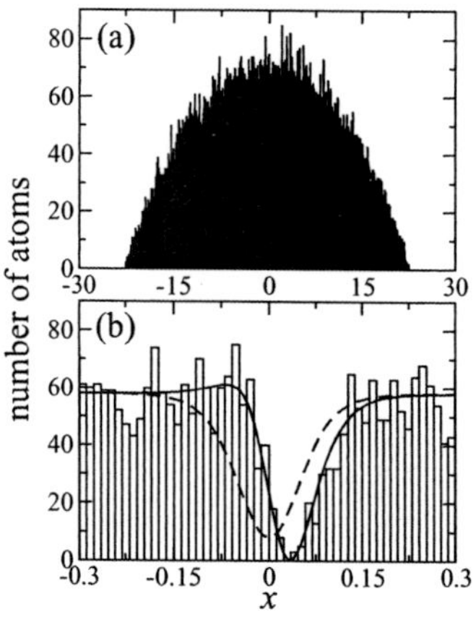

Figure 6: Density measurement for state $|0\rangle_b$, Eq.(70). Panel (a) shows the whole atomic cloud while panel (b) the region around the trap center only. The histogram represents number of particles in bins measured in a single experiment. The solid line stands for $A\phi_1^2(x) + B\phi_+^2(x) + C\phi_1(x)\phi_+(x)$, where A, B, C are parameters fitted to the histogram. The dashed line gives the expected average outcome of many experiments i.e. the single particle density (73). The values of the parameters correspond to the Hannover experiment [4], where $g = 7500$ and $N = 1.5 \times 10^5$.

dark soliton shown in Fig. 6. In Fig. 7 we show how definite values of $\cos\varphi$ and q are established in the course of measurement of subsequent atomic positions. It is enough to measure only a small fraction of atoms to suppress fluctuations of φ and q, see Fig. 7. Figure 6 shows that soliton in a single experiment appears to be dark but generally it is displaced from the center of the trap. Because the localization of the soliton in each experiment is random, averaging atomic densities over many experiments leads to a gray soliton localized at the trap center, i.e. exactly what is predicted by the single particle density function.

In current experiments [4] the dark solitons are generated with the help of the phase imprinting method [5]. As we argued, when the depletion in the condensate before the imprinting is negligible, then the soliton state right after imprinting can be idealized by the state without any atoms depleted from the solitonic condensate ϕ_1 to the anomalous mode ϕ_+,

$$|\psi(t=0)\rangle = |N, 0\rangle \ . \tag{77}$$

This state is not a stationary state. To obtain time evolution of this state we consequently truncate the Bogoliubov Hamiltonian to the anomalous mode where

$$H_B \ = \ -\frac{1}{\sqrt{2}}\,\hat{b}^\dagger\hat{b} \tag{78}$$

with the quasi-particle annihilation operator $\hat{b}$ that in the number-conserving Bogoliubov

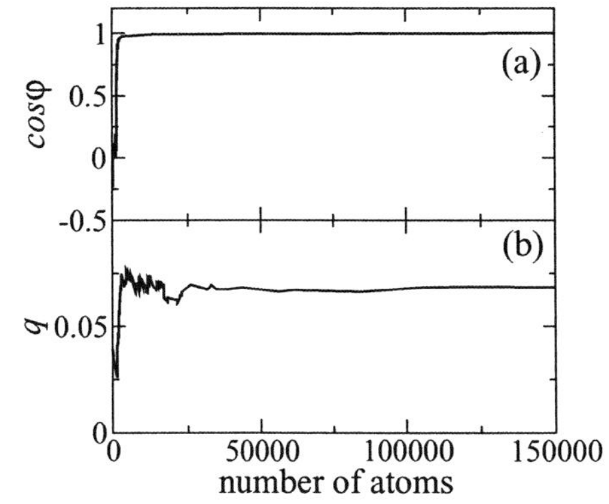

Figure 7: Values of $\cos\varphi$ [panel (a)] and q [panel (b)] during the density measurement performed on the state $|0\rangle_b$. After the position of each atom is determined and the atom is annihilated from the state, see (59) and the corresponding text, we calculate eigenstates of the single particle density matrix which allows us to determine the condensate fraction and the condensate wave function. The latter is always of the form (74) and that allows us to obtain values of $\cos\varphi$ and q.

theory reads

$$\hat{b} = \frac{\hat{a}_1^\dagger \hat{a}_+ - \lambda_+ \hat{a}_1 \hat{a}_+^\dagger}{\sqrt{N(1-\lambda_+^2)}} . \tag{79}$$

With this Hamiltonian the initial perfect condensate $|\psi(t=0)\rangle$ evolves into a depleted state $|\psi(t)\rangle$. In Fig. 8 we show how a single particle density evolves in time. After about 15 ms (for the parameters of the Hannover experiment [4]) the hole in the single particle density distribution fills up with depleted atoms. We also simulated single realizations of destructive density measurements of the state $|\psi(t)\rangle$. A generic density distribution after evolution for 10 ms is shown in Fig. 9. We see that the soliton is dark and it is displaced with respect to the trap center. Inspection of the state after 10 ms shows that

$$\langle N-k, k|\psi(10\,\mathrm{ms})\rangle \sim e^{ik\varphi} \tag{80}$$

with a tiny $\varphi \simeq 0.01$. In principle the time evolution might generate any phase $\varphi \in [0, 2\pi)$, but the parameters of the experiment [4] are such that there is not enough time to get any more significant φ from the idealized initial state (77) before the linearized Bogoliubov theory breaks down.

In a realistic experiment the dark soliton is created in neither the stationary nor the zero depletion state. The two states may appear "natural" to a theorist's eye accustomed to such

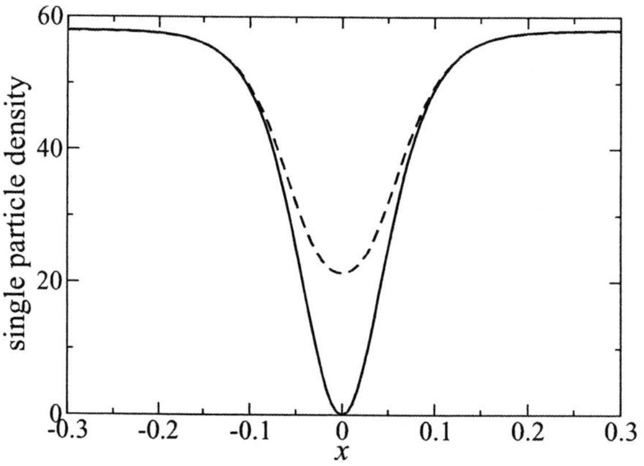

Figure 8: Solid line stands for single particle density of the state $|\psi(0)\rangle$, Eq.(77). Dashed line is single particle density for $|\psi(t = 10\text{ms})\rangle$. The values of the parameters correspond to the Hannover experiment [4], where $g = 7500$ and $N = 1.5 \times 10^5$.

highly idealized examples. However, these states are not realistic from experimental point of view. We do not know what the actual state, or ensemble of states, is. What we can do is to test how robust is the picture of the condensate soliton obtained for the Bogoliubov vacuum and for the evolving initially perfect condensate with respect to imperfections of the quantum state.

To get an idea what are images of a generic non-stationary state we model the state by a state with random phases. The Hamiltonian of the system is diagonal in the Fock basis of quasi-particles and not in the particle Fock basis. We expect a generic non-stationary state to have Fock representation with phases that appear random. To be more specific, we take as an example the Bogoliubov vacuum (70) but with randomized phases α_k,

$$|\text{rand}\rangle \sim \sum_{k=0}^{N/2} e^{i\alpha_k} \lambda^k \frac{\sqrt{(N-2k)!(2k)!}}{(N/2-k)!k!} |N-2k, 2k\rangle \; . \tag{81}$$

Single particle density in this state is exactly the same as for the Bogoliubov vacuum, see Eq. (73), but possible outcomes of single density measurements can be very different in the two states. Indeed, a generic outcome in the state $|\text{rand}\rangle$ shown in Fig. 10 reveals a displaced gray soliton with $\cos\varphi = 0.45$. The phase disorder present in the state $|\text{rand}\rangle$ translates into random φ's.

As the last example, we take the state $|\text{rand}\rangle$ as an initial state instead of the state (77), evolve it for 0.7 ms, and then measure atomic positions. In Fig.11 we show a single particle density after 0.7 ms and a corresponding generic outcome of density measurement. Not surprisingly, we again see a displaced gray soliton, this time with $\cos\varphi = -0.48$.

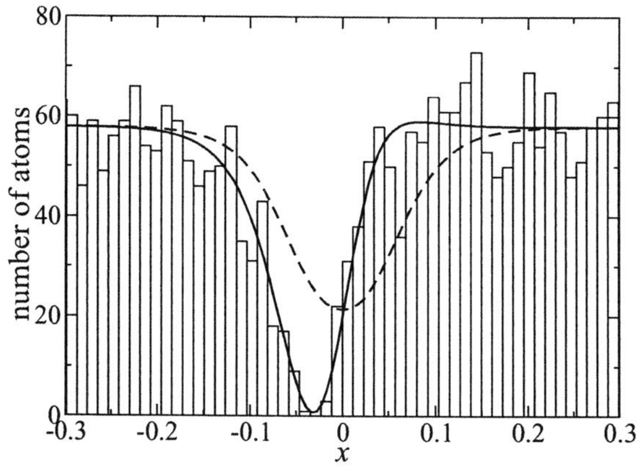

Figure 9: Density measurement for state $|\psi(t = 10\text{ms})\rangle$, Eq. (77). The histogram represents number of particles in bins measured in a single experiment. The solid line stands for $A\phi_1^2(x) + B\phi_+^2(x) + C\phi_1(x)\phi_+(x)$, where A, B, C are parameters fitted to the histogram. The dashed line gives the expected average outcome of many experiments. The values of the parameters correspond to the Hannover experiment [4], where $g = 7500$ and $N = 1.5 \times 10^5$.

The last two examples demonstrate that phase disorder in particle Fock representation can change qualitatively possible results of density measurements.

5 Conclusion

The dark soliton is a solution of the one dimensional Gross-Pitaevskii equation. However, this non-linear Schrödinger equation is only a mean field description of a Bose-Einstein condensate [3]. The description takes into account interactions between atoms but only on a mean-field level. Any depletion of atoms from the condensate wave function by two-body collisions is neglected. The mean field approximation not only neglects the so called quantum depletion in the ground state of the condensate — which in fact in many situations is quite small — but also any depletion that might arise in a dynamical situation when the system is prepared in a non-stationary state.

We used both the traditional symmetry breaking version of the Bogoliubov theory [8] and the number-conserving Bogoliubov theory [10, 14] to analyze depletion of atoms from a condensate with a dark soliton. In the mean field approximation the notch of the condensate wave function with the dark soliton is empty. It turns out that in the stationary Bogoliubov vacuum state the notch is partially filled with atoms depleted from the condensate. Density of depleted atoms in the center of the notch is 14% of the total density of the condensate for experimental parameters [4]. This quantum depletion is quite substantial and indicates

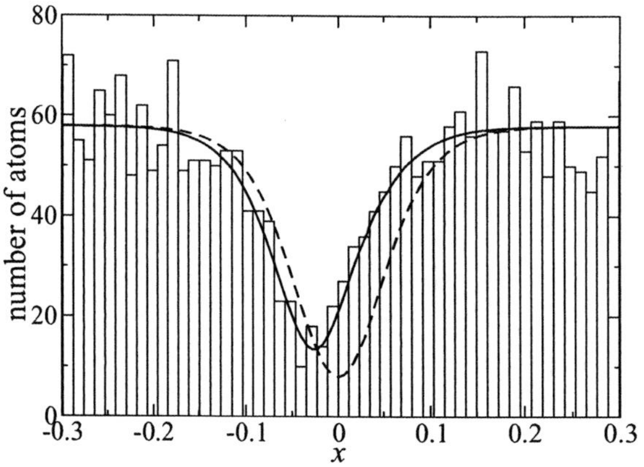

Figure 10: The same as Fig. 9 but for the state $|\mathrm{rand}\rangle$, Eq. (81).

that in the presence of the soliton depletion effects are strongly amplified as compared to the ground state of the condensate without any soliton.

We also used the Bogoliubov theory in a dynamical situation to describe evolution of atomic depletion after the soliton is generated by the phase imprinting method [4, 5]. In this non-stationary situation initial depletion is much less than the 14% but after about 10 ms the notch of the dark soliton disappears filled up with depleted atoms. The 10 ms are in a striking coincidence with the time dark solitons need to loose contrast in the Hannover experiment [4]. The mechanism behind this number is quite robust, as it involves only the most fundamental effect of atomic collisions.

The single particle density calculated in the Bogoliubov theory indicates that the notch in the dark soliton fills up with atoms. This density is a reduced single particle density which is an *expectation* value of the density operator in the quantum state of N atoms. As usual in quantum mechanics the quantum state does not tell us where the atoms are but only where they are *expected* to be. To find out where they actually are we have to measure where they are but this measurement destroys the original quantum state. From the work of Javanainen and Yoo [12] we know that in a Bose-condensed system a destructive density measurement can give outcomes which are qualitatively different from the average density distribution. In this way we arrive at a non-trivial question: averaging over many experimental realizations we will see a picture where the soliton is filled up with depleted atoms but what shall we actually see when we measure the density of atoms in a single experiment. Shall we see an empty dark soliton with a soliton position that fluctuates from experiment to experiment or a gray soliton filled with a random amount of atoms. On average both extremes give the same average density of atoms.

We employed the number-conserving Bogoliubov theory to construct explicit N-particle atomic states and then simulated destructive density measurements in these states. We found

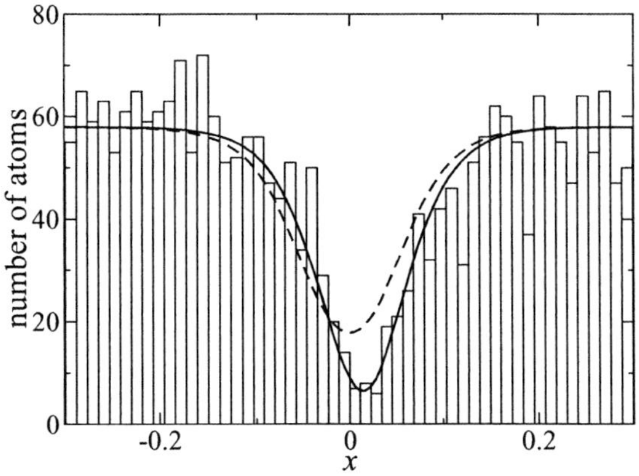

Figure 11: The same as Fig. 9 but for the state $|\mathrm{rand}\rangle$, Eq. (81), after evolution over 0.7 ms.

that in a highly idealized states like the Bogoliubov vacuum state or the state without initial depletion the density measurement finds the dark soliton empty but displaced from the origin to a random position. However, in examples of more generic states the soliton is found both displaced from the origin and filled with a random amount of atoms. In a generic state the density measurement reveals that the dark soliton is loosing contrast due to atomic depletion.

APPENDIX

5.1 The N-particle State of BEC

Here we briefly summarize main conclusions of the number-conserving Bogoliubov theory of a Bose-Einstein condensate [10, 14]. A field operator $\hat{\psi}(\vec{r})$ is split into a condensate part $\hat{a}_0\Phi_0(\vec{r})$ and a non-condensate part $\delta\hat{\psi}(\vec{r})$,

$$\hat{\psi}(\vec{r}) = \Phi_0(\vec{r})\hat{a}_0 + \delta\hat{\psi}(\vec{r}) \ . \tag{82}$$

To zero order in the small fluctuation operator $\delta\hat{\psi}(\vec{r})$ the N-particle state is a perfect condensate

$$\left(\hat{a}_0^\dagger\right)^N |0\rangle \tag{83}$$

with all N atoms in the condensate wave function $\Phi_0(\vec{r})$. The Φ_0 (normalized to unity) solves a stationary Gross-Pitaevskii equation

$$\mu\Phi_0 = -\frac{\hbar^2}{2m}\nabla^2\Phi_0 + V(\vec{r})\Phi_0 + g_0 N|\Phi_0^2|\Phi_0 \ . \tag{84}$$

To zero order the interaction affects only the shape of the condensate wave function $\Phi_0(\vec{r})$ through the nonlinear term $g_0 N|\Phi_0^2|\Phi_0$ in the Gross-Pitaevskii equation.

Quantum depletion from this condensate shows up in the second order of the perturbation theory where the N-particle state becomes a pair-correlated Bogoliubov vacuum state

$$|0\rangle_b \sim \left(d^\dagger\right)^{N/2}|0\rangle = \left(\hat{a}_0^\dagger\hat{a}_0^\dagger + \sum_{s=1}^{\infty}\lambda_s\hat{a}_s^\dagger\hat{a}_s^\dagger\right)^{N/2}|0\rangle. \tag{85}$$

The sum runs over an orthonormal basis of non-condensate modes $\Phi_s(\vec{r})$ orthogonal to $\Phi_0(\vec{r})$. This general pair-correlated ansatz (85) has been known for a while, see e.g. the review by Leggett [15], but a general solution for the coefficients λ_s and corresponding wave functions Φ_s has been found only very recently in Ref.[14] (the solution in the special case of a uniform condensate was given in Ref.[15]).

The construction of the Bogoliubov vacuum (85) runs in the following steps — its full justification can be found in Refs.[10, 14]. First we solve the stationary Gross-Pitaevskii equation (84) to get $\Phi_0(\vec{r})$ and μ. Note that Φ_0 does not need to be the ground state, it can be an excited stationary state with a dark soliton or vortex. Then we solve the linear Bogoliubov-de Gennes equations for the Bogoliubov normal modes of small fluctuations around Φ_0,

$$\begin{aligned}
&-\frac{\hbar^2}{2m}\nabla^2 U_m + V(\vec{r})U_m + 2g_0N|\Phi_0|^2U_m + g_0N\Phi_0^2V_m = \\
&\mu U_m + \omega_m U_m\ , \\
&-\frac{\hbar^2}{2m}\nabla^2 V_m + V(\vec{r})V_m + 2g_0N|\Phi_0|^2V_m + g_0N(\Phi_0^*)^2U_m = \\
&\mu V_m - \omega_m V_m
\end{aligned} \tag{86}$$

to get the eigenvalues ω_m and Bogoliubov eigenmodes $U_m(\vec{r}), V_m(\vec{r})$ corresponding to the original Bogoliubov theory. The U_m, V_m must be projected on the subspace orthogonal to Φ_0,

$$\begin{aligned}
u_m &= U_m - \Phi_0\langle\Phi_0|U_m\rangle\ , \\
v_m &= V_m - \Phi_0\langle\Phi_0|V_m\rangle\ ,
\end{aligned} \tag{87}$$

in order to get the Bogoliubov eigenmodes (u_m, v_m) relevant in the number-conserving Bogoliubov theory. After the projection the modes (u_m, v_m) are normalized so that a norm

$$\langle u_m|u_m\rangle - \langle v_m|v_m\rangle = 1\ . \tag{88}$$

All modes with negative norm are ignored. The normalized modes define bosonic quasi-particle annihilation operators

$$\hat{b}_m = \frac{\hat{a}_0^\dagger}{\sqrt{N}}\langle u_m|\delta\hat{\psi}\rangle - \frac{\hat{a}_0}{\sqrt{N}}\left(\langle v_m^*|\delta\hat{\psi}\rangle\right)^\dagger \tag{89}$$

which transfer atoms between the condensate and the non-condensate modes, but conserve the total number of atoms N. For small depletion the Hamiltonian can be approximated by a sum of harmonic oscillators

$$\hat{H} = \sum_m \omega_m \, \hat{b}^\dagger_m \hat{b}_m \; . \tag{90}$$

The N-particle Bogoliubov vacuum (85) is the eigenstate of $\hat{H}$ without any quasi-particles,

$$\hat{b}_m|0\rangle_b = 0 \; . \tag{91}$$

The N-particle state (85) is annihilated by all $\hat{b}$'s provided $\hat{d}^\dagger$ defined in Eq.(85) commutes with all $\hat{b}$'s,

$$\left[\hat{b}_m, \hat{d}^\dagger\right] = 0 \; . \tag{92}$$

To construct such a $\hat{d}^\dagger$ operator one has to choose an orthonormal basis of states $\tilde{\Phi}_s(\vec{r})$ orthogonal to Φ_0, and then calculate matrices $\tilde{U}$ and $\tilde{V}$ with elements

$$\tilde{U}_{mk} = \langle u_m|\tilde{\Phi}_s\rangle \; , \tag{93}$$

$$\tilde{V}_{mk} = \langle\tilde{\Phi}_s|v_m\rangle \; . \tag{94}$$

Employing the matrices $\tilde{U}$ and $\tilde{V}$ allows us to write Eqs. (92) in the form

$$\tilde{V} = \tilde{U}\tilde{Z} \tag{95}$$

Inverting $\tilde{U}$ we obtain matrix $\tilde{Z}$

$$\tilde{Z} = \tilde{U}^{-1}\,\tilde{V} \; . \tag{96}$$

Diagonalization of $\tilde{Z}$ results in desired eigenvalues λ_s and orthonormal eigenmodes Φ_s. Having calculated λ_s and Φ_s we can construct the operator $\hat{d}^\dagger$ and the Bogoliubov vacuum state (85). Excited eigenstates can be obtained by repeated action of the quasi-particle creation operators $\hat{b}^\dagger_m$ on the vacuum state.

Acknowledgments

We acknowledge the important role of Z. P. Karkuszewski in obtaining some of our results reviewed here and for discussions. We acknowledge also helpful discussions with James Anglin, Diego Dalvit, Yvan Castin, and Jakub Zakrzewski. This research was supported by the Polish Ministry of Scientific Research Grant PBZ-MIN-008/P03/2003.

References

[1] J. K. Taylor, Ed., *Optical Solitons Theory & Experiment* (Cambridge Univ. Press, New York, 1992).

[2] M. H. Anderson, J. R. Ensher, M. R. Matthews, C. E. Wieman, and E. A. Cornell, *Science* **269**, 198 (1995); K. B. Davis, M.-O. Mewes, M. R. Andrews, N. J. van Druten, D. S. Durfee, D. M. Kurn, and W. Ketterle, *Phys. Rev. Lett.* **75**, 3969 (1995); C. C. Bradley, C. A. Sackett, and R. G. Hulet, *Phys. Rev. Lett.* **78**, 985 (1997); for a review see *Bose-Einstein Condensation in Atomic Gases, Proceedings of the International School of Physics "Enrico Fermi"*, edited by M. Inguscio, S. Stringari, and C. Wieman (IOS Press, Amsterdam, 1999).

[3] L.P. Pitaevskii, *Sov. Phys. JETP* **13**, 451 (1961); E.P. Gross, *Nuovo Cimento* **20**, 454 (1961); F. Dalfovo, S. Giorgini, L. P. Pitaevskii, and S. Stringari, *Rev. Mod. Phys.* **71**, 463 (1999).

[4] S. Burger K. Bongs, S. Dettmer, W. Ertmer, K. Sengstock, A. Sanpera, G. V. Shlyapnikov, and M. Lewenstein, *Phys. Rev. Lett.* **83**, 5198 (1999); J. Denschlag, J. E. Simsarian, D. L. Feder, C. W. Clark, L. A. Collins, J. Cubizolles, L. Deng, E. W. Hagley, K. Helmerson, W. P. Reinhardt, and S. L. Rolston and B. I. Schneider W. D. Phillips, *Science* **287**, 97 (2000).

[5] Ł. Dobrek, M. Gajda, M. Lewenstein, K. Sengstock, G. Birkl, and W. Ertmer, *Phys. Rev. A* **60**, R3381 (1999).

[6] A. Muryshev, G.V. Shlyapnikov, W. Ertmer, K. Sengstock, and M. Lewenstein. *Phys. Rev. Lett.* **89**, 110401 (2002).

[7] J. Dziarmaga, Z.P. Karkuszewski, and K. Sacha, *Phys. Rev. A* **66**, 043615 (2002); J. Dziarmaga and K. Sacha, *Phys. Rev. A* **66**, 043620 (2002); J. Dziarmaga, Z.P. Karkuszewski, and K. Sacha, *J. Phys. B* **36**, 1217 (2003).

[8] Ph. Nozieres and D. Pines, *The Theory of Quantum Liquids*, (Addison Wesley, New York, 1990), Vol. II; A.L. Fetter, *Ann. Phys.* (N.Y.), **70**, 67 (1972).

[9] A. E. Muryshev, H. B. van Linden van den Heuvell, and G. V. Shlyapnikov, *Phys. Rev. A* **60**, R2665 (1999); P. O. Fedichev, A. E. Muryshev, G. V. Shlyapnikov, *Phys. Rev. A* **60**, 3220 (1999); Th. Busch and J. R. Anglin, *Phys. Rev. Lett.* **84**, 2298 (2000).

[10] M. D. Girardeau and R. Arnowitt, *Phys. Rev.* **113**, 755 (1959); C. W. Gardiner, *Phys. Rev. A* **56**, 1414 (1997); M. D. Girardeau *Phys. Rev. A* **58**, 775 (1998); Y. Castin and R. Dum, *Phys. Rev. A* **57**, 3008 (1998); Y. Castin, in *Les Houches Session LXXII, Coherent atomic matter waves 1999*, edited by R. Kaiser, C. Westbrook and F. David, (Springer-Verlag Berlin Heilderberg New York 2001).

[11] We estimate the one-dimensional g coefficient by $4\pi NaL/S$ where a is the s-wave scattering length, $L = \sqrt{\hbar/m\Omega_x}$, and S is an average transverse area calculated within the Thomas-Fermi approximation.

[12] J. Javanainen and S.M. Yoo, *Phys. Rev. Lett.* **76**, 161 (1996).

[13] M. Naraschewski et al., *Phys. Rev. A* **54**, 2185 (1996); Y. Castin and J. Dalibard, *Phys. Rev. A* **55**, 4330 (1997); R.A. Hegstrom, *Chem. Phys. Lett.* **288**, 248 (1998); K. Mølmer, *Phys. Rev. A* **65**, 021607 (2002); S. Ashhab and A.J. Leggett, *Phys. Rev. A* **65**, 023604 (2002).

[14] J. Dziarmaga and K. Sacha, *Phys. Rev. A* **67**, 033608 (2003).

[15] A.J. Leggett, *Rev. Mod. Phys.* **73**, 307 (2001).

In: Advances in Soliton Research
Editor: L.V. Chen, pp. 135-151

ISBN 1-59454-769-6

Chapter 4

SOLITONS ON NANOTUBES AND FULLERENES AS SOLUTIONS OF A MODIFIED NON-LINEAR SCHRÖDINGER EQUATION

Yves Brihaye **and** ***Betti Hartmann*** [†]
*Faculté des Sciences, Université de Mons-Hainaut, 7000 Mons, Belgium
[†]School of Engineering and Sciences, International University Bremen (IUB), 28725 Bremen, Germany

Abstract

Fullerenes and nanotubes consist of a large number of carbon atoms sitting on the sites of a regular lattice. For pratical reasons it is often useful to approximate the equations on this lattice in terms of the continuous equation. At the moment, the best candidate for such an equation is the modified non-linear Schrödinger equation. In this paper, we study the modified non-linear Schrödinger equation, which arises as continuous equation in a system describing an excitation on a hexagonal lattice. This latter system can e.g. describe electron-phonon interaction on fullerene related structures such as the buckminster fullerene and nanotubes. Solutions of this modified non-linear Schrödinger equation, which have solitonic character, can be constructed using an Ansatz for the wave function, which has time and azimuthal angle dependence introduced previously in the study of spinning boson stars and Q-balls. We study these solutions on a sphere, a disc and on a cylinder.

Our construction suggests that non-spinning as well as spinning solutions, which have a wave function with an arbitrary number of nodes, exist. We find that this property is closely related to the series of well-known mathematical functions, namely the Legendre functions when studying the equation on a sphere, the Bessel functions when studying the equation on a disc and the trigonometric functions, respectively, when studying the equation on a cylinder.

PACS numbers: 73.61.Wp, 11.27.+d

*E-mail address: Yves.Brihaye@umh.ac.be
[†]E-mail address: b.hartmann@iu-bremen.de

1 Introduction

Fullerenes and fullerene related structures have gained a lot of interest in the past 20 years. Fullerenes were discovered for the first time in 1985 [1] and are carbon-cage molecules with a large number n of carbon (C) atoms bonded in a nearly spherically symmetric configuration. The configuration is such that three of the valence electrons of the carbon atom form the bounds with the neighbouring carbon atoms. This means, that one "free" valence electron can "hop" along the the n positions in the fullerene.

The C_{60} fullerene, the so-called "buckminster" fullerene, has 60 carbon atoms arranged on a lattice with 20 hexagons and 12 pentagons. It has a diameter of $d_{60} \approx 7$ Å. In general, fullerenes with n carbon atoms consists of lattices with 12 pentagons and $(\frac{n}{2}-10)$ hexagons, where $n \geq 24$ has to be even.

Fullerides are alkali-doped fullerenes, i.e. fullerenes on which alkali metal atoms such as rubidium (Rb), potassium (K), caesium (Cs) or sodium (Na) are put. These alkali metals donate one electron each. It was observed that such "doping" leads to a metallic or even superconducting behaviour [2]. The interesting thing about this is that the transition temperature T_c of the fullerides is quite high for superconductors, e.g. $T_c = 33$ K for a $RbCs_2C_{60}$ [3]. In an attempt to explain this high transition temperature, it was found that phonon-electron interactions in the fullerides can explain this phenomenon quite well [4].

Nanotubes are fullerene related structures, which were discovered for the first time in 1991 [5]. These molecules have cylindrical shape and the carbon atoms are arranged on a hexagonal grid. The ends of the nanotubes are typically closed by caps of pentagonal rings. While the tubes discovered in 1991 were multi-layered, single-walled tubes were synthesised in 1993 [6]. The tubes are distinguished by their chirality, diameter and lengths. They have average diameters of 1.2 to 1.4 nm and can be up to 2 mm long. Concerning their chirality, the tubes are typically put into three different classes: armchair, zigzag or chiral tubes [7]. Depending on their structure, the tubes have different mechanical, thermal, optical and electrical properties. The chirality determines e.g. whether the tubes behave metallic or semiconducting. These facts also lead to the conclusion that a distortion of the hexagonal lattice would influence the energy-band gap. A distortion can be achieved in two ways: either by external, mechanical forces such as bending, stretching or twisting or by an interaction of the lattice with an internal excitation. Trying to explain the dispersion-free energy transport in biopolymers, Davydov realised [8] that the interaction of an internal excitation with the distorted lattice (whose distortion was initially caused by the internal excitation) leads to a localised state, a soliton [9]. For the case of biopolymers, the internal excitation is an amide I vibration.

Motivated by the research on fullerene related structures, the interaction of a 2-dimensional hexagonal lattice with an excitation caused by an "excess electron" was studied recently [10]. The Hamiltonian is of Fröhlich-type and describes "electron-phonon" interactions. It was found that the existence of soliton-like structures depends crucially on the value of the "electron-phonon" coupling.

Interestingly, in the continuum limit, this latter system of equations reduces to a modified non-linear Schrödinger equation, where the extra term appearing due to the discreteness of the lattice and the "electron-phonon" interaction, leads to the stabilisation of the soliton. The appearance of this type of equation has been observed also for a similar system on a quadratic lattice [11, 12]. Consequently, the modified non-linear Schrödinger equation was studied on a 2-dimensional plane in [13] and on a sphere in [14] using an Ansatz previously introduced in the study of boson stars [15] and Q-balls [16]. Non-spinning as well as spinning solutions with and without nodes were constructed.

In this paper, we study the modified non-linear Schrödinger equation arising as continuum limit of the equations describing the interaction of an electron-like excitation with a hexagonal lattice. In Section II, we review some generalities. Using the Ansatz introduced in [15, 16], we construct solutions for three different cases, which we describe in Sections III-V. In Section III, we discuss the results for the solutions on the sphere obtained in [14] from another point of view, namely we give the spectrum of the equation. In Section IV, we discuss the solutions on a 2-dimensional disc with fixed radius and demonstrate how the solutions obtained on the plane [13] can be understood in this context. In Section V, we discuss the equation on a cylinder with fixed radius and fixed length. We give our conclusions in Section VI.

2 The Modified Non-linear Schrödinger Equation: Generalities

The starting point is a coupled systems of discrete equations describing the interaction of an excitation (e.g. an "excess electron") with a regular lattice in two dimensions. The excitation is given in terms of a complex valued scalar field ψ, while the lattice displacements are described by two fields, one standing for the displacement in x, the other for that in y direction. These discrete equations were studied for the quadratic lattice in [11, 12] and for the hexagonal lattice in [10]. Remarkably, it was found for the hexagonal lattice that in the stationary limit it is possible to replace the full system of equation, in which the excitation field ψ and the displacement fields are coupled, by a modified discrete non-linear Schrödinger equation. This is not possible for the quadratic lattice.

In the continuum limit, both systems of equations reduce to a modified non-linear Schrödinger equation. The new term appearing in this equation, which results from both the discreteness of the lattice as well as from the interaction between the lattice and the complex valued scalar field ψ ("lattice-excitation interaction"), leads to the stabilisation of the solution. The difference between the quadratic and the hexagonal case is just the coefficient in front of the non-linear term in the equation, which, however is of the same order of magnitude for the two cases.

Since we are dealing with fullerene related structures here, we consider the modified non-linear Schrödinger equation, which arises as continuum limit in the hexagonal case [10, 13]. In dimensionless variables it reads:

$$i\frac{\partial\psi}{\partial t} + \Delta\psi + 4g\psi\left(|\psi|^2 + \frac{1}{8}\Delta|\psi|^2\right) = 0 \,. \tag{1}$$

g is the coupling constant of the system and determines the "strength" of the non-linear character of the equation. Especially, as can be seen from the derivation of the modified non-linear Schrödinger equation from the discrete equations [10], the mass of the carbon atom is encoded into g.

3 Equation on the Sphere

In this section, we will discuss the equation (1) above on a 2-dimensional sphere of radius R_s aiming to describe "electronic" excitations on fullerenes and fullerides.

The Laplacian operator in this case reads:

$$\Delta_{sphere} = \frac{1}{R_s^2}\left(\frac{\partial^2}{\partial\theta^2} + \cot\theta\frac{\partial}{\partial\theta} + \frac{1}{\sin^2\theta}\frac{\partial^2}{\partial\varphi^2}\right) , \tag{2}$$

where $\theta \in [0 : \pi]$ and $\varphi \in [0 : 2\pi]$. Note that $\psi = \psi(t, \theta, \varphi)$ is a complex valued scalar field.

Solutions of (1) considered on the sphere with radius R_s can be characterised by their norm:

$$\eta^2 = \int_0^{2\pi}\int_0^{\pi} |\psi|^2 \sin\theta d\theta d\varphi \tag{3}$$

as well as by their energy:

$$E = \int_0^{2\pi}\int_0^{\pi} \left(|\vec{\nabla}\psi|^2 - 2g|\psi|^4 + \frac{1}{4}g(\vec{\nabla}|\psi|^2)^2\right) \sin\theta d\theta d\varphi . \tag{4}$$

We will construct normalised solutions, i.e. $\eta^2 = 1$. Note that in [14] a different strategy was employed. By rescaling the wave-function $\psi \rightarrow \psi/\sqrt{g}$, the coupling constant g can be eliminated from the equation. (Of course, the normalisation of the wave-function is lost in this approach, but since η was plotted, the corresponding value of g could be easily determined.) In this paper, we put the emphasis on the evolution of the spectrum (corresponding to normalised eigenfunctions) as a function of the coupling constant g. Note that g determines the "strength" of the non-linear part of the equation and that for $g = 0$, the equation becomes the ordinary Schrödinger equation on the sphere, which, of course, is linear.

3.1 Discussion of the Solutions

To construct explicit solutions, we employ the following axially symmetric Ansatz [15, 16]:

$$\psi(t, \theta, \varphi) = e^{-i\omega t + im\varphi}\Phi(\theta) . \tag{5}$$

The solutions with $m = 0$ are non-spinning solutions, while the spinning solutions will have $m \neq 0$.

Inserting this Ansatz into (1), we find the following equation [14]:

$$\Phi'' + \cot\theta\Phi' + \Phi'^2\frac{g\Phi}{1+g\Phi^2} + \frac{4g\Phi^3+\omega\Phi}{1+g\Phi^2}R_s^2 - \frac{m^2}{1+g\Phi^2}\frac{1}{\sin^2\theta}\Phi = 0\,, \tag{6}$$

where the prime denotes the derivative with respect to θ.

This is a priori a non-linear equation with external parameters g and ω. Here, we will consider it from a slightly different point of view. Let us rewrite (6) as follows:

$$-\left(\Phi'' + \cot\theta\Phi' + \Phi'^2\frac{g\Phi}{1+g\Phi^2} + \frac{g\Phi^3R_s^2(4-\omega)}{1+g\Phi^2} - \frac{m^2}{1+g\Phi^2}\frac{1}{\sin^2\theta}\Phi\right) = \Omega\Phi \tag{7}$$

where

$$\Omega \equiv \omega R_s^2 \tag{8}$$

is the effective spectral parameter. The equation then appears as an eigenvalue Schödinger equation supplemented by a non-linear term coupled with the parameter g. The solutions in the linear case, i.e. for $g=0$ are well-known and it is very likely that - by continuity- their deformation due to the non-linear term can be constructed at least for small values of g. However, this does not exclude the occurence of supplementary branches of solutions, which may appear typically due to the non-linear structure of the equation.

In order to construct regular solutions, (7) has to be solved subject to the following boundary conditions:

$$\Phi'|_{\theta=0} = \Phi'|_{\theta=\pi} = 0 \quad \text{for} \;\; m=0 \tag{9}$$

and

$$\Phi(\theta=0) = \Phi(\theta=\pi) = 0 \quad \text{for} \;\; m\neq 0\,. \tag{10}$$

To solve the equation (7) numerically subject to the boundary conditions (9), respectively (10) for generic values of g and R_s, we use a numerical routine based on the Newton-Raphson method [17]. Our strategy is to use $\Phi(\theta=0)$, respectively $\Phi'|_{\theta=0}$ as a "shooting" parameter for the case $m=0$ and $m\neq 0$.

3.1.1 The Case $g=0$

In the case $g=0$, (7) reduces to the well-known Legendre equation

$$\Phi'' + \cot\theta\Phi' - m^2\frac{1}{\sin^2\theta}\Phi + \Omega\Phi = 0\,. \tag{11}$$

The solutions of this equation are the associated Legendre functions:

$$\Phi(\theta) = P_l^m(\cos\theta) \quad \text{with} \;\; \Omega = l(l+1)\,, \tag{12}$$

where

$$l\in\mathbb{N}\;,\quad m=-l,-l+1,...,l-1,l\,. \tag{13}$$

The spatial part of (5) are then the spherical harmonics $Y^l_m(\theta,\varphi) = e^{im\varphi}P^l_m(\cos\theta)$. Notice, however, that in the equation studied here there is no distinction between m and $-m$. From now on, we will assume m to be positive, but it is understood that the solution with corresponding $-m$ can be trivially obtained.

The spherical harmonics have a well-defined parity, i.e. they are either symmetric or anti-symmetric under the reflection $\Phi(\theta) \to \Phi(\pi-\theta)$. In addition, the number of zeros in the interval $]0:\pi[$ is equal to $l-|m|$.

These results are, of course, very well- known, but we mention them for completeness and because the spectrum of the full equation will approach the spectrum of the Legendre equation in the $g \to 0$ limit.

3.1.2 Deformed Legendre Functions

All the solutions discussed in the previous section are continuously deformed when the parameter g is chosen to be larger than zero. For every available spherical harmonic, we observe that a branch of solutions in g exists, on which the solutions have the same symmetry than in the $g=0$ limit. Our numerical results further indicate that the number of zeros k of the solution is preserved all along the branch and is equal to $k \equiv l-m$.

Now, we will discuss in detail the numerical solutions corresponding to the deformations of the spherical harmonics Y^0_0, Y^1_0 and Y^1_1.

- The non-spinning ($m=0$) and nodeless ($l=0$, $k=0$) solution corresponding to the case Y^0_0 is in fact the constant solution of the equation, namely

$$\Phi = \sqrt{\frac{-\omega}{4g}} \ . \tag{14}$$

  The eigenvalue Ω of the corresponding normalised solution has a linear dependence of the parameter g:

$$\Omega = -\frac{R_s^2 g}{\pi} \ . \tag{15}$$

  This dependence is shown in Fig.1 for $R_s=2$.

- The non-spinning ($m=0$), one-node ($l=1$, $k=1$) solution is anti-symmetric under the reflection $\theta \to \pi-\theta$. The function $Y^1_0 \propto \cos\theta$ is progressively deformed by the non-linear term for $g>0$. For $g \approx 10$ the profile of the function $\Phi(\theta)$ possesses a plateau $\Phi=0$ surrounding the region of $\theta=\pi/2$. Away from this plateau, the solutions resemble two disconnected solitons located at the two poles of the sphere, i.e. the maxima of $\Phi^2(\theta)$ are situated around $\theta=\pi$ and $\theta=0$. Stated in other words, the probability density vanishes in a region surrounding the equator, while it has local maxima at the two poles. Let us stress, that the physical parameters remain monotonic functions of the "shooting" parameter $\Phi(\theta=0)$.

The dependence of Ω on g is shown in Fig.1 for $R_s = 2$. For $g = 0$, we have $\Omega = 2$, as expected. It is worth noticing that the two eigenvalues corresponding to the quantum numbers $m = l = 0$ and $m = 0$, $l = 1$ cross at $g \approx 1.5$. This would a priori mean, that the "excited" solution becomes the "ground state" of the equation. However, as we will see in the next section, the appearance of new asymmetric solutions will turn out to have even lower values of Ω.

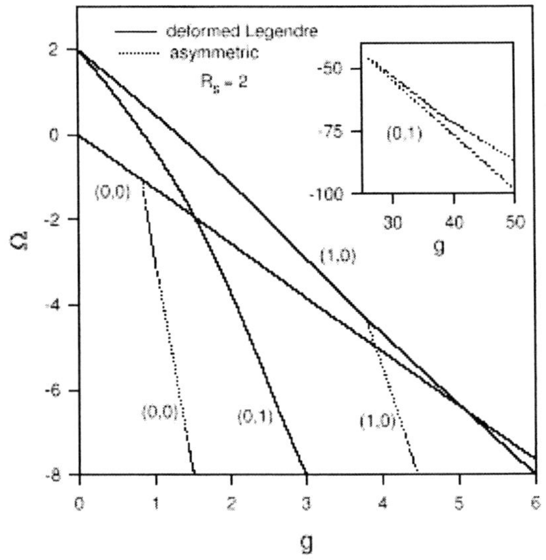

Figure 1: The eigenvalue $\Omega = \omega R_s^2$ is shown as a function of g for $(m, l) = (0, 0)$, $(0, 1)$ and $(1, 0)$. We have chosen $R_s = 2$. The solid lines denote the deformed Legendre functions, while the dotted lines are the asymmetric solutions (see text for more details).

- The spinning ($m = 1$), no-node ($l = 1$, $k = 0$) solutions arrive as deformations of the spherical harmonic $Y_1^1 \propto \sin\theta e^{i\varphi}$. This solution is thus symmetric under the reflection $\theta \to \pi - \theta$. In contrast to the non-spinning solutions, we notice here that when the coupling constant g is of order 2.5, the structure of the solutions becomes very sensitive to the value of the "shooting" parameter $\Phi'|_{\theta=0}$. In other words, there exists a critical value of $\Phi'|_{\theta=0} \equiv \Phi'_{(cr)}|_{\theta=0}$ (depending on R_s) such that no solutions exist for $\Phi'|_{\theta=0} > \Phi'_{(cr)}|_{\theta=0}$. Our numerical results, however, suggest that solutions of arbitrary large values of g can be constructed when considering the limit $\Phi'|_{\theta=0} \to \Phi'_{(cr)}|_{\theta=0}$. The corresponding eigenvalue of Ω decreases monotonically with g from $\Omega = 2$ (for the case $g = 0$) as can be seen in Fig.1 for the case $R_s = 2$. Of course, this curve also represents the solution with $m = -1$, so that the eigenvalue Ω is three-fold degenerated in the limit $g = 0$.

Finally, let us state that for these solutions, the function Φ has a maximum at the equator and vanishes at the poles. This is an interesting result with view to [18]. In this latter paper it was found that if a C_{60} or a C_{70} molecule is doped with one or two excess electrons, the additional charges accumulate nearly along an equatorial line of the molecule.

Of course, deformations of the Legendre functions for higher m and l can be constructed in a systematic way.

3.1.3 Asymmetric Solutions

All solutions of the non-linear equation (7) discussed previously appear as continuous deformations of the solutions available in the linear limit $g = 0$. However, in this section, we will exhibit new branches of solutions, which are fundamentally non-linear phenomena and as these have no linear counterparts. Numerically, we find that these new branches of solutions exist for sufficiently high values of the coupling constant g parametrising the non-linear effect and that some of these branches bifurcate into the branches of the deformed Legendre functions at critical values of $g = g_{cr}^{(m,l)}(R_s)$, which depend on l, m (and thus on the number of nodes $k = l - m$) as well as on the radius R_s.

These solutions are charaterised by the fact that they are asymmetric with respect to the reflection $\theta \to \pi - \theta$.

- Asymmetric, non-spinning ($m = 0$) and nodeless ($l = 0, k = 0$) solutions exist for $g > g_{cr}^{(0,0)}(R_s = 2) \approx 0.838$. When the parameter g is chosen larger than this critical value, our numerical results indicate that this solution has a maximum at $\theta = 0$ and a minimum at $\theta = \pi$ and decreases monotonically between these two extrema, which are strictly positive. Thus, the solution is neither symmetric nor anti-symmetric under the reflection $\theta \to \pi - \theta$. Instead, this reflection applied to the solution leads to an equivalent solution with a maximum (respectively minimum) at $\theta = \pi$ ($\theta = 0$). The solutions thus represent a physical state with probability density concentrated in one of the hemispheres. This is a new feature that cannot arise in the deformation of the Legendre functions.

  In the limit $g \to g_{cr}^{(0,0)}$ the values of the maximum and the minimum approach each other and the solution becomes constant. The branch of these asymmetric solutions bifurcates into the branch of constant solutions (deformed Y_0^0) described in the previous section. The corresponding value of ω of these asymmetric solutions is lower than the one of the symmetric branch. This is shown in Fig.1 (dotted line) for $R_s = 2$.

  For generic values of R_s, the bifurcation of the asymmetric branch occurs at

$$g_{cr}^{(0,0)}(R_s) = \frac{\pi}{R_s^2 - \frac{1}{4}} , \tag{16}$$

with

$$\Omega_{cr}^{(0,0)} = -\frac{R_s^2}{R_s^2 - 1/4} . \tag{17}$$

- Asymmetric, non-spinning ($m = 0$) and one-node ($l = 1$, $k = 1$) solutions exist for $g > g_{cr}^{(0,1)}(R_s = 2) \approx 25$. In this region of parameter values, we find that in fact two branches of solutions exist (see Fig.1). For a fixed value of g, the solution with the lower value of Ω has a bigger value of $\Phi(0)$ and its minimum lies at smaller values of θ. This is demonstrated for $g = 37$, $R_s = 2$ in Fig.2. The values of ω corresponding to these asymmetric solutions are higher than the values of the corresponding (i.e. with for the same values of g) antisymmetric solution on the branch of deformed Legendre functions Y_0^1. In contrast to the other described cases, we have not succeeded to construct another (a third) branch which would eventually bifurcate from the branch of deformed Legendre functions.

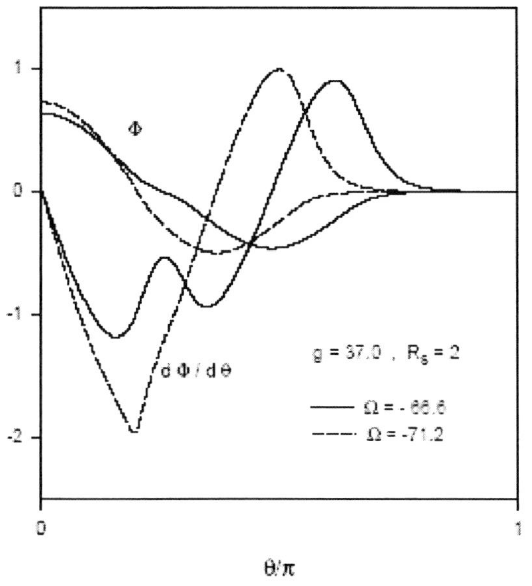

Figure 2: The profiles of the function $\Phi(\theta)$ and the derivative $\frac{d\Phi}{d\theta}$ are shown for $(m, l) = (0, 1)$ (non-spinning, nodeless solutions), $g = 37$, sphere-radius $R_s = 2$ and two different values of the spectral parameter $\Omega = \omega R_s$.

- Asymmetric, spinning ($m = 1$) and nodeless ($l = 1$, $k = 0$) solutions exist for $g > g_{cr}^{(1,1)}(R_s = 2) \approx 3.79$. Solving the equations for values of g larger than this critical value, we were able to construct a branch of asymmetric solutions (and the corresponding mirror symmetric one). The maximum of these solutions is located between $\theta = 0$ and $\theta = \pi/2$ such that the probability density is located in one of the hemispheres. As before, the frequency of the asymmetric solution is lower than the frequency of the corresponding (i.e. same g) solution on the deformed Y_1^1 branch.

The construction of asymmetric branches with higher values of the spin m and/or higher node numbers k becomes very tricky. This is mainly due to the occurence of many solutions existing for the same values of the parameters and also due to the fact that the profiles of the symmetric and asymmetric (or anti-symmetric and asymmetric) solutions can be very close to each other on a large part of the $[0:\pi]$ interval. The construction of asymmetric branches can thus not be achieved systematically.

4 Equation on the Disc

In [13] the equation (1) was discussed on a two-dimensional plane and non-spinning as well as spinning solutions with nodes were constructed. Here, we will treat this problem differently. With view to the discussion of the equation on a sphere, we will discuss here the equation on a two-dimensional disc with radius R_d parametrized by the coordinates ρ and φ. The case studied in [13] then arises as limit $R_d \to \infty$. The Laplacian is given by:

$$\Delta_{disc} = \frac{\partial^2}{\partial\rho^2} + \frac{1}{\rho}\frac{\partial}{\partial\rho} + \frac{1}{\rho^2}\frac{\partial^2}{\partial\varphi^2} \tag{18}$$

with $\varphi \in [0:2\pi]$. The norm reads:

$$\eta^2 = \int_0^{2\pi}\int_0^{R_d} |\psi|^2 \rho d\rho d\varphi \tag{19}$$

and the energy is given by:

$$E = \int_0^{2\pi}\int_0^{R_d} \left(|\vec{\nabla}\psi|^2 - 2g|\psi|^4 + \frac{1}{4}g(\vec{\nabla}|\psi|^2)^2\right) \rho d\rho d\varphi \; . \tag{20}$$

4.1 Discussion of the Solutions

As for the case on the sphere, we adopt the following Ansatz for the complex valued function $\psi(t,\rho,\varphi)$:

$$\psi(t,\rho,\varphi) = e^{-i\omega t + im\varphi}\phi(\rho) \; . \tag{21}$$

Inserting this Ansatz gives the following equation:

$$\phi'' + \frac{\phi'}{\rho} + \frac{g\phi\phi'^2}{1+g\phi^2} + \frac{4g\phi^3 + \omega\phi}{1+g\phi^2} - \frac{m^2\phi}{\rho^2(1+g\phi^2)} = 0 \; , \tag{22}$$

where the prime now denotes the derivative with respect to ρ.

We can rewrite, analogue to before, the equation in the following way:

$$-\left(\phi'' + \frac{\phi'}{\rho} + \frac{g\phi\phi'^2}{1+g\phi^2} + \frac{g\phi^3(4-\omega)}{1+g\phi^2} - \frac{m^2\phi}{\rho^2(1+g\phi^2)}\right) = \omega\phi \; . \tag{23}$$

This equation looks like the one studied in [13], however, we impose different boundary conditions here:

$$\partial_\rho\phi|_{\rho=0} = 0 \;\; , \;\; \phi(R_d) = 0 \;\; \text{for} \;\; m = 0 \tag{24}$$

and

$$\phi(\rho=0)=0 \ , \ \ \phi(R_d)=0 \ \ \text{for} \ \ m \neq 0 \ . \tag{25}$$

Again, we used the numerical routine described in [17] to solve the equation (23) subject to the boundary conditions (24), respectively (25).

4.1.1 The $g = 0$ Limit

In the limit $g = 0$ (23) reduces to the linear Schrödinger equation in polar coordinates

$$\phi'' + \frac{\phi'}{\rho} - \frac{m^2\phi}{\rho^2} + \omega\phi = 0 \ . \tag{26}$$

Note that for static solutions, i.e. $\omega = 0$, the equation reduces to the Euler equation with solutions of the form $\psi(t,\rho,\varphi) \equiv \psi(\rho) \propto e^{im\varphi}\rho^m$. For $\omega \neq 0$, the equation becomes the Bessel equation if we define $x \equiv \sqrt{|\omega|}\rho$ and $F(x) \equiv \phi(x)$:

$$\frac{d^2F}{dx^2} + \frac{1}{x}\frac{dF}{dx} + \left(1 - \frac{m^2}{x^2}\right)F = 0 \tag{27}$$

The solutions of this equation, which are regular at the origin are of course the well-known Bessel functions $J_m(x)$ for $\omega > 0$ and the modified Bessel functions $I_m(x)$ for $\omega < 0$. Remembering that the Bessel functions $J_m(x)$ are oscillating and admit an infinite number of nodes on the positive real axis, it is easy to construct solutions of (26) with k nodes. This is possible assuming that

$$\phi(\rho) = J_m(\sqrt{\omega}\rho) \ \ \text{with} \ \ \phi(R_d) = 0 \tag{28}$$

and choosing the value of ω in such the way that the value $\rho = R_d$ coincides with the k-th zero (we call it $x = x_{m,k}$) of the Bessel function J_m, i.e. $\sqrt{\omega} = x_{m,k}/R_d$.

This indicates that in the limit $g = 0$ solutions with angular momentum m and with k nodes exist and the spectrum of the equation is positive. The corresponding functions are convex.

4.1.2 The $g \neq 0$ Case

If the coupling constant g is positive but small, we expect that the pattern of solutions will be similar to the one available in the limit $g = 0$ discussed above. Our numerical results indeed confirm this expectation. For small values of g, the solutions are still represented by convex functions and the values of ω are positive and decrease slowly with g as shown in Fig. 3 for the case $m = 0$, $k = 0$. Let is call this range of g-values the "weak coupling-regime".

When g is large enough, the value of ω becomes negative (see Fig.3) and the solutions have an inflexion point at some intermediate value of ρ (see Fig.4). As a consequence they become much more localized in the region around the origin. In this "strong coupling-regime" the solutions can thus be described as being soliton-like. We also note that the value of ω decreases very strongly with g.

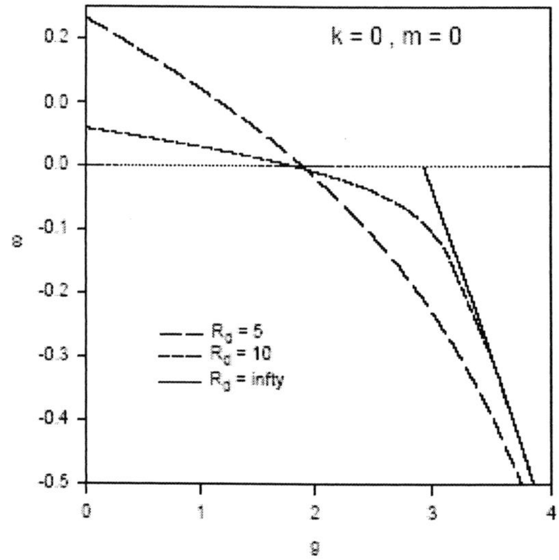

Figure 3: The eigenvalue ω is shown as a function of g for the non-spinning ($m = 0$), nodeless ($k = 0$) solutions on the disc. We have chosen $R_d = 5$, 10 and ∞. Note that $R_d = \infty$ corresponds to the case of the plane.

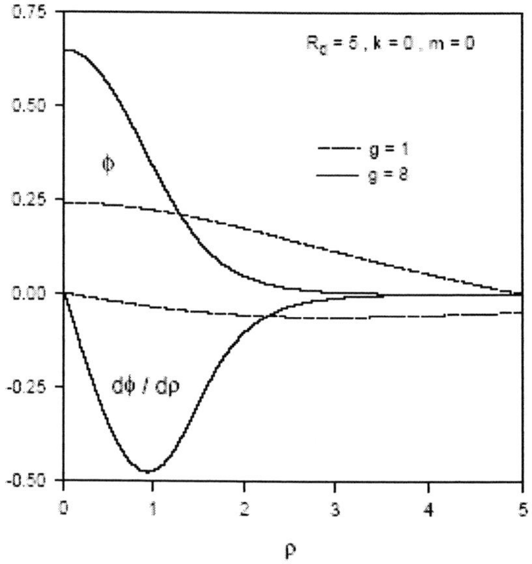

Figure 4: The profiles of the function $\phi(\rho)$ and the derivative $\frac{d\phi}{d\rho}$ are shown for non-spinning ($m = 0$), nodeless ($k = 0$) solutions on the disc with radius $R_d = 5$ for two different values of g.

The difference between the strong and weak coupling is shown in Fig.4 for $g = 1$, respectively $g = 8$ and $R_d = 5$. The value ω is shown in dependence on g for the non-spinning ($m = 0$) and nodeless ($k = 0$) solution for different values of the disc-radius R_d in Fig.3.

In Fig. 3 we also superposed the values of ω corresponding to the case of the full plane (i.e. $R_d = \infty$). As has been shown in [10, 13], in this case, solutions exists only for large enough values of g: $g \geq g_{cr} \approx 2.94$. It is therefore natural to try to understand the pattern available on the plane as the $R_d \rightarrow \infty$ limit of the pattern of solutions for the disc. In this respect, it is interesting to notice that in the "strong coupling-regime" the values of ω corresponding to a finite disc get very close (especially for the case corresponding to $R_d = 10$) to the values of ω associated with the case of the infinite disc. This is related to the fact that with increasing g, the "soliton" becomes more and more peaked around the origin and doesn't "notice" whether the disc is finite or infinite. This observation gives a natural explanation for the fact that solutions exist only for $g \geq g_{cr}$ in the case of the plane. Indeed in the case of the plane, it was noticed [13] that the linearized equation (valid asymptotically) also is of Bessel type. However, the nodeless, localized solutions are exponentially decreasing and can only be associated with the modified Bessel function, which correspond to negative values of ω. That's why, on the plane, nodeless soliton-like solutions exist only for $g \geq 2.94$ and have no "weak coupling"-counterparts. For large values of g the numerical results strongly suggest the following linear relation between g and the frequency corresponding to the fundamental mode : $\omega \approx -0.66g + 2.19$. We also studied the solutions with $k > 0$ nodes and $m > 0$ and found that the pattern is similar to the case $k = 0$ and $m = 0$ discussed above in detail.

We do not find it useful to discuss them here, but we believe that our results provide a robust evidence that regular solutions of (23) with an arbitrary number of nodes and arbitrary (integer) angular momentum exist on the disc/plane.

5 Equation on the Cylinder

We also considered the equation on a cylinder of radius R_c and length $2L$. This is a natural extension of the previous results in order to describe soliton-like structures on nanotubes with radius R_c and length $2L$. Using the natural coordinates z, φ, the Laplacian reads:

$$\Delta_{cylinder} = \frac{1}{R_c^2}\frac{\partial^2}{\partial\varphi^2} + \frac{\partial^2}{\partial z^2} \tag{29}$$

with $\varphi \in [0 : 2\pi]$ and $z \in [-L, L]$. The norm reads:

$$\eta^2 = \int_0^{2\pi}\int_{-L}^{L} |\psi|^2 dz d\varphi \tag{30}$$

and the energy is given by:

$$E = \int_0^{2\pi} \int_{-L}^{L} \left(|\vec{\nabla}\psi|^2 - 2g|\psi|^4 + \frac{1}{4} g (\vec{\nabla}|\psi|^2)^2 \right) dz d\varphi \ . \tag{31}$$

5.1 Discussion of the Solutions

We use the following Ansatz:

$$\psi(t, z, \varphi) = e^{-i\omega t + im\varphi} F(z) \ , \quad m = 0, 1, 2, \ldots \tag{32}$$

The equation (1) then becomes:

$$F'' + (F')^2 \frac{gF}{1+gF^2} + \left(4gF^2 + \omega - \frac{m^2}{R_c^2} \right) \frac{F}{1+gF^2} = 0 \ , \tag{33}$$

where the prime denotes the derivative with respect to z.

Again, we can rewrite this equation in terms of a "spectral equation":

$$- \left[F'' + (F')^2 \frac{gF}{1+gF^2} + \left(gF^2(4-\omega) - \frac{m^2}{R_c^2} \right) \frac{F}{1+gF^2} \right] = \omega F \ . \tag{34}$$

This equation has to be solved subject to the following boundary conditions:

$$F(-L) = F(L) = 0 \ . \tag{35}$$

The equation (33) is then solved subject to the boundary conditions (35) using the numerical routine described in [17].

5.1.1 The Case $g = 0$

The case $g = 0$ corresponds to an harmonic equation:

$$F'' = - \left(\omega - \frac{m^2}{R_c^2} \right) F \ . \tag{36}$$

The solutions of this equation subject to the boundary conditions (35) are:

$$F(z) = \sin\left((z - L)\sqrt{\omega - m^2/R_c^2} \right) \quad \text{with} \quad \omega = \left(\frac{m}{R_c} \right)^2 + \left(\frac{\pi n}{2L} \right)^2 \ , \quad n = 1, 2, 3, \ldots \tag{37}$$

5.2 The Case $g \neq 0$

It is possible to solve (33) by quadrature, which gives:

$$(F')^2 = \frac{C - 2gF^4 - \Omega F^2}{1 + gF^2} \quad , \quad \Omega \equiv \omega - \frac{m^2}{R_c^2} \tag{38}$$

and C is an integration constant. The integration can be performed, however, it leads to an involved and untractable expression of z in terms of F, C, Ω. Assuming the cylinder to be infinite ($L = \infty$) and $F(\infty) = F'(\infty) = 0$ leads to $C = 0$. If we now require in addition that the solution is symmetric under the reflexion $z \rightarrow -z$ (i.e. $F'(0) = 0$) we obtain the relation $\Omega = -2gF(0)^2$. This provides a useful check for our numerical results.

Performing a numerical analysis of the values of ω as a function of g (with L, R_c, k, m fixed), we obtain a pattern very similar to that available for the case on a disc (see Fig.3). The only noticable difference resides in the fact that, in the limit $L = \infty$, a normalizable solution can be constructed for all values of the parameter g. This case is, of course, interesting with iew to nanotubes since the lengths $2L$ of a tube is much larger than its diameter $2R_c$. In the case of the "ground state" solution, which is non-spinning ($m = 0$) and nodeless ($k = 0$), the following relations hold :

$$\omega \approx -0.1g^2 \ \text{for} \ g \ll 1 \quad , \quad \omega \approx -0.555g + 1.28 \ \text{for} \ g \gg 1 \,. \tag{39}$$

Unlike in the case of the sphere, we did not succeed in constructing asymmetric solutions in the case of the cylinder (i.e. solutions, which are neither even nor odd under the reflexion $z \rightarrow -z$) and we believe that there are no such solutions.

(40)

6 Conclusions

In this paper we have presented an extended analysis of a modified non-linear Schrödinger equation in 1+2 dimensions. This equation was constructed as an effective continuous equation describing an interaction of an excitation with a 2-dimensional hexagonal lattice. In this paper, we think of the excitation as an "excess electron" described by a wave function ψ and the interaction to be "electron-phonon" -like. As is well-known from amide I-excitations in biopolymeric lattices such an interaction leads to the creation of a localised state, a "soliton".

The continuous modified non-linear Schrödinger equation was obtained with the idea to approximate the physical system on the plane [10, 13], but it can be modified in order to be considered on different geometrical manifolds, which are justified physically. These are, namely, a cylinder to mimick excitations on a carbon nanotube and a sphere, respectively, in order to describe excitations (e.g. extra electrons) on a fullerene nanomolecule, which is particularly interesting with view to the possibility of describing the high transition temperature of superconducting fullerides.

Completed with the different types of boundary conditions related to the geometry and using an Ansatz previously introduced in the study of boson stars [15] and spinnning Q-balls [16], the equation (1) possesses a lot symmetries, some are continuous (like rotations), some are discrete (like reflexions).

In all cases considered the spectrum of solutions is rich, many "normal-mode"-types of solutions exist, mainly characterized by two integers: the internal angular momentum (or spin) and the number of nodes of the wave function. This is very reminiscent to the principal quantum numbers of more familiar quantum mechanical systems.

It might turn out to be difficult to classify the solutions of a non-linear, partial differential equations. The various types of solutions, however, can be traced back from the corresponding solutions available in the weak coupling limit, in which the equation becomes linear and its spectrum is known analytically. In the case of the sphere, disc and cylinder, respectively, we found deformed Legendre functions, deformed Bessel functions and deformed harmonic functions. However, in the case of the sphere, we demonstrated the existence of additional solutions, which are specifically related to the non-linear character of the equation. The main feature distinguishing these supplementary solutions is their asymmetry with respect to the reflexion at the equator of the sphere. In other words, we exhibited the spontaneous symmetry breaking (SSB) of one of the natural symmetries of the initial problem.

This SSB occurs when the non-linear coupling becomes strong enough. Similarly it can occur for very different types of non-linear equations. As an example, we point out the bifurcation of bisphaleron solutions from the sphaleron solutions available in the case of the classical equations of the electroweak field theory [19, 20]. In this case, the parity operator is spontaneously broken by the so called bisphaleron solutions.

Acknowledgments

Y.B. gratefully acknowledges the Belgian F.N.R.S. for financial support.

References

[1] H. W. Kroto, J. R. Heath, S. C. O'Brien, R. F. Curl and R. E. Smalley, *Nature* **318** (1985), 162.

[2] R. C. Haddon *et al.*, *Nature* **350** (1991), 320; *for a review see* O. Gunnarsson, *Rev. Mod. Phys.* **69** (1997), 575.

[3] K. Tanigaki, T. W. Ebbesen, S. Saito, J. Mizuki, J. S. Tsai, Y. Kubo and S. Kuroshima, *Nature* **352** (1991), 222.

[4] C. M. Varma, J. Zaanen and K. Raghavaclari, *Science* **254** (1991), 989; M. A. Schluter, M. Lannoo, M. Needles, G. A. Baraff and D. Tomanek, *Phys. Rev. Lett.* **68** (1992), 526; *J. Phys. Chem. Solids* **53** (1992), 1473; I. I. Mazin, S. N. Rashkeev,

V. P. Antropov, O. Jepsen, A. I. Lichtenstein anbd O.K. Andersen, *Phys. Rev.* **B45** (1992), 5114.

[5] S. Iijima, *Nature* (London) **354** (1991), 56.

[6] S. Iijima and T. Ichihashi, *Nature* (London), **363** (1993), 603.

[7] see e.g. M. S. Dresselhaus, G. Dresselhaus and P. Eklund, *The Science of fullerenes and carbon nanotubes* (Academic, New York, 1996); *Carbon nanotubes, Preparation and Properties*, edited by T. W. Ebbesen (CRC, Boca Raton, FL, 1996); R. Saito, G. Dresselhaus and M. S. Dresselhaus, *Physical Properties of carbon nanotubes* (World Scientific, Singapore, 1998); P. J. F. Harris, *Carbon Nanotubes and Related Structures* (Cambridge University Press, Cambridge, 1999); *Carbon nanotubes: Synthesis, Structure, Properties and Applications*, edited by M. S. Dresselhaus, G. Dresselhaus and P. Avouris (Springer Verlag, Berlin, 2000).

[8] A. S. Davydov, *Solitons in Molecular systems* (Reidel, Dordrecht, 1985).

[9] A. Scott, Phys. Rep. **217** (1992), 1; *Nonlinear excitations in Biomolecules*, edited by M. Peyrard (Springer, Berlin, 1996).

[10] B. Hartmann and W. J. Zakrzewski, *Phys. Rev.* **B 68** (2003), 184302.

[11] L. Brizhik, A. Eremko, B. Piette and W. J. Zakrzewski, *Physica* **D 146** (2000), 275; **D 159** (2001), 71.

[12] L. Brizhik, B. Piette and W. J. Zakrzewski, *Ukr. Fiz. Zh.* (Russ. Ed.) **46** (2001), 503.

[13] Y. Brihaye, B. Hartmann and W. J. Zakrzewski, *Phys. Rev.* **D 69** (2004), 087701.

[14] Y. Brihaye and B. Hartmann, *J. Phys.* **A 37** (2004), 1181.

[15] F. E. Schunck and E. W. Mielke, in *Relativity and Scientific Computing*, Springer, Berlin (1996), 138.

[16] M. Volkov and E. Wöhnert, *Phys. Rev.* **D 66** (2002), 085003.

[17] U. Ascher, J. Christiansen and R. D. Russell, *Math. Comput.* **33**, (1979), 659; *ACM Trans. Math. Softw.* **7** (1981), 209.

[18] K. Harigaya, *Phys. Rev.* **B45** (1992), 13676.

[19] J. Kunz and Y. Brihaye, *Phys. Lett.* **B 216**, 353 (1989).

[20] L. G. Yaffe, *Phys. Rev.* **D 40**, 2723 (1989).

In: Advances in Soliton Research
Editor: L.V. Chen, pp. 153-186
ISBN 1-59454-769-6

Chapter 5

The Cascade is a MMS Instanton

Jarah Evslin*
Physique Théorique et Mathématique, Université Libre
de Bruxelles, C.P. 231, B-1050, Bruxelles, Belgium

Abstract

Wrap m D5-branes around the 2-cycle of a conifold, place n D3-branes at a point and watch the system relax. The D5-branes source m units of RR 3-form flux on the 3-cycle, which cause dielectric NS5-branes to nucleate and repeatedly sweep out the 3-cycle, each time gaining m units of D3-charge while the stack of D5-branes loses m units of D3-charge. A similar description of the Klebanov-Strassler cascade has been proposed by Kachru et al. when $m >> m - n$, where it is a tunneling event in the dual field theory. Using the T-dual MQCD we argue that the above process occurs for any m and n and in particular may continue for more than one step. The nonbaryonic roots of the SQCD vacua lead to new cascades because, for example, the 3-cycle swept does not link all of the D5's. This decay is the S-dual of a MMS instanton, which is the decay into flux of a brane that is trivial in twisted K-theory. This provides the first evidence for the S-dual of the K-theory classification that does not itself rely upon any strong/weak duality.

1 Introduction

The reader interested in the cascade but not the K-theory classification may skip to Subsec. 2.3.

Given a compactification of type II string theory, how can we classify the topologically stable configurations? If we are only interested in configurations without branes, then we need to classify fluxes, which are differential forms. Classically this requires considering fluxes that solve the equations of motion modulo the applicable gauge invariances, quantum

*E-mail address: jarah@df.unipi.it

mechanically one may also need to apply a quantization condition and include some torsion corrections to the equations of motion. Considering the equations of motion and gauge invariances

$$dG_{p+1} = H \wedge G_{p-1}, \qquad C_p \longrightarrow C_p + d\Lambda_{p-1} + H \wedge \Lambda_{p-3} \tag{1.1}$$

of the Ramond-Ramond fluxes defined below one finds, as is reviewed in section 2, that after including torsion corrections RR fluxes are classified by twisted K-theory [1, 2, 3].

This construction realizes twisted K-theory as a quotient (gauge orbits) of a subset (solutions of equations of motion) of integral cohomology. One may try to apply the same logic to conclude that D-branes are classified by twisted K-homology, which is a quotient of a subset of integral homology. However in the case of D-branes the states that should be excluded in fact are legitimate solitons, they are the baryons of Ref. [4]. In addition the D-brane configurations that should be equivalent are not quite equivalent. There are dynamical processes, called MMS instantons, that turn a configuration into another that represents the same K-homology class [5]. While successfully interpolating between homologically distinct D-branes, these processes also change the background fluxes. The new flux can have dramatic physical effects. For example it can make a type IIA theory massive [5, 6] or, as is the subject of this paper, it can be the H flux produced during the Klebanov-Strassler [7] cascade and thus contribute to the superpotential of the corresponding $N = 1$ gauge theory. In the case of the D3-brane decay described in the abstract the resulting flux implies that for each D3-brane created by an NS5-brane a D5-brane loses a unit of D3-charge.

The twisted K-theory classification of the RR fluxes faces similar problems. The fluxes corresponding to cohomology classes that are not K-theory classes are configurations with branes, and so rather than being unphysical, like the baryons they merely lie outside the set of configurations considered. Meanwhile the field strengths that are trivial in twisted K-theory may be gauged away, but this transformation leaves a residual integral, globally defined gauge field, which is a higher dimensional analog of a theta angle that is a multiple of 2π. While this may seem trivial, if we allow NS5-branes to sweep out cycles in our spacetime until H integrated over every cycle vanishes, any flux that seemed to have been gauged away will be restored[1] [8]. Thus the theta angles remember the original flux configuration, and so the fluxes corresponding to trivial twisted K-theory classes cannot in fact be pure gauge.

Even if we ignore these problems by considering only configurations without branes and identifying configurations with different theta angles, RR fluxes are not classified by twisted K-theory. In addition to the equations of motion and Bianchi identities (1.1) for the RR fields which yield the K-theory classification, one must also impose the equations of motion and quotient by the Bianchi identity of the NS 3-form field strength $H = dB$ and its dual $*H = dB^{(6)}$. In type IIB string theory, for example these are easily found by

[1]This is a result of the Chern-Simons supergravity coupling of the theta angle to H and the restored field strength. In the quantum theory this process, if possible, must continue until $W_3 + H = 0$ on each cycle, and correspondingly the gauge transformations must be augmented by Steenrod square terms so that the same cancellation occurs.

S-duality to be

$$d * H = G_3 \wedge G_5, \qquad B \longrightarrow B + d\Lambda_1, \qquad B^{(6)} \longrightarrow B^{(6)} + d\Lambda_5 + G_3 \wedge \Lambda_3 \quad (1.2)$$

where Λ_3 is the same 3-form used to simultaneously transform C_4. It is tempting to S-dualize $dG_3 = H \wedge G_1$ to find dH, however when the right hand side is nontrivial the usual arguments for S-duality fail, and indeed one appears to find that the dilaton is not globally defined. While such bizarre configurations do appear, for example, in F-theory on K3 [9], it seems impossible to disentangle them from the topology of the ambient spacetime and so they apparently defy the kind of classification scheme with which we are concerned here. Luckily no such flux appears in the Klebanov-Strassler geometry.

The extra restrictions and quotients of Eq. (1.2) imply that RR fluxes are classified by something very different from twisted K-theory. In fact, it has been known for some time [10] that the K-theory classification is not consistent with S-duality in type IIB string theory. Thus the classification of branes in type IIB must be some S-duality covariant modification of K-theory, where the above problems with the twisted K-theory classification are hidden by excluding configurations with baryons, identifying configurations with different fluxes and cutting out any NS5-branes that source these fluxes, or in the S-dual version of interest cutting out the D5-brane sources. The usual construction of K-theory from vector bundles does not appear to admit an obvious covariantization. A more promising construction, by Maldacena, Moore and Seiberg (MMS) in Ref. [5], uses physical processes, called MMS instantons. While they consider instantons constructed entirely from D-branes, the construction appears to work with fundamental strings and NS5-branes as well and so, modulo problems with 7-branes discussed above, is easily made S-duality covariant. Using this S-duality covariantized MMS construction a proposed S-duality covariant classification of branes and strings in type IIB was proposed in Refs. [11, 8].

The only nontrivial test of this proposal to date has been by Loaiza-Brito in Ref. [12], who has shown that it is consistent with the $SL(2,\mathbb{Z})$ duality of 4-dimensional $N = 4$ supersymmetric gauge theories. In this note we provide a second test. D3-branes at a point on the deformed conifold with m units of G_3 flux are classified by $\mathbb{Z}$ in ordinary twisted K-theory, and so should never decay. However in the S-duality covariant generalization, and even in the S-dual of the ordinary K-theory classification, they are classified by $\mathbb{Z}_m$, and so m D3-branes may decay at a time. The cascade is precisely a series of such decays, and so it is consistent with the S-duality covariant K-theory classification and not with the ordinary K-theory classification. Furthermore the MMS instanton that corresponds to this decay is the decay described at large m by Kachru, Pearson and Verlinde (KPV) in Ref. [13], where our use of the T-dual MQCD eliminates the need to introduce auxilliary anti-branes to break the supersymmetry and so drive this process[2]. As described above, such instantons leave a residual flux. In this case there is one unit of H flux for every m D3-branes that decay. Using the T-dual MQCD we are able to extend these results to small m and n, where the usual arguments for the deformation of the conifold do not apply. We will find that D-branes are still classified by the (S-dual) twisted K-theory of the deformed conifold.

[2]The process is driven by the tension of the M5-brane, which is lost in the 10d $g_s = 0$ dimensional reduction.

We begin in Sec. 2 with a review of the twisted K-theory classification of fluxes and branes. We argue that Page field strengths correspond to the integral cohomology classes of the Atiyah-Hirzebruch spectral sequence while gauge-invariant field strengths are the twisted Chern characters [14, 15] of the twisted K-theory classes. In Sec. 3 we recall that the $SU(2)$ WZW model also exhibits a cascade and show that this cascade is U-dual to a process similar to that of KPV and valid at all m and n. In Sec. 4 we review the Klebanov-Strassler cascade in MQCD, where we can see features not readily apparent in the conifold description. In addition to the baryonic vacua of the theory studied by Klebanov and Strassler [7], one also sees the nonbaryonic vacua which lead to different cascades. We consider generalizations of the 4d gauge theory, such as $N = 2$ softly broken to $N = 1$ by a superpotential that is a nondegenerate polynomial in the adjoint chiral multiplet, where one finds copies of the same story located at each extremum of the superpotential. Finally in Sec. 5 we T-dualize this description to the conifold, where we find the KPV process.

2 The Twisted K-Theory Classification

2.1 Classifying Field Strengths

If C_p are the RR gauge fields and B the NS two-form then there are at least two kinds of RR field strengths that we may be interested in classifying [16]. The first is the gauge invariant

$$G_{p+1} = dC_p - H \wedge C_{p-2} \tag{2.1}$$

where $H = dB$ is the NS 3-form field strength. The Bianchi identity $ddC_p = 0$, which is equivalent to the equation of motion in Eq. (1.1), implies that the gauge invariant field strengths are $(d - H)$ closed. We will say that a configuration is topologically trivial[3] if the q-form connections C_q are all globally defined, in which case these field strengths are $(d - H)$ exact. Thus the gauge invariant field strengths G_p are classified by H-twisted cohomology, which is the cohomology with respect to the differential operator $d - H$.

The second is the Page field strength $F_{p+1} = dC_p$, which classically is classified by de Rham cohomology as a result of the Bianchi identity $dF_{p+1} = 0$ and the fact that C_p is globally well-defined if F_{p+1} is exact. The well-definedness of Dp-brane partition functions leads to a Dirac quantization condition[4], and so F_p's must represent integral cohomology classes. Eq. (1.1) together with $H \wedge H = 0$ implies the descent relation

$$H \wedge F_{p-1} = dG_{p+1}. \tag{2.2}$$

As G_{p+1} is gauge invariant, the right hand side is exact and so cohomologically trivial. Therefore F_{p-1} belongs to a special subset of cohomology that is annihilated by H. Further

[3] The K-theory classification is not sensitive to the generalized theta angles.

[4] Although the derivative of the D-brane partition function actually contains the gauge invariant field strength G_{p+1}, a bulk contribution [17] to the anomaly cancels the extra term leaving the anomaly equal to $exp(i \int (F_{p+1}))$.

if F_{p+1} is equal to H wedged with a globally defined, closed $(p-2)$-form ω_{p-2} then F_{p+1} may be gauged away by the patchwise transformation

$$C_p \longrightarrow C_p + H \wedge \eta_{p-3}, \qquad d\eta_{p-3} = \omega_{p-2} \tag{2.3}$$

where η_{p-3} exists on each patch because ω_{p-2} is closed. We have then learned that the Page field strengths correspond to those elements of integral cohomology in the kernel of $H\cup$ quotiented by those in its image, up to torsion corrections that are not apparent in a supergravity analysis.

If we assume that the only possible field strengths on cycles are those that could be generated by filling in those cycles with a contractible space and including branes that source the flux, then we find a possible form for such torsion corrections from the Freed-Witten [19] anomaly on the branes. Instead of taking the cohomology with respect to H, we must take it with respect to $Sq^3 + H$ and then with respect to an unknown degree five operator[5]. This is the Atiyah-Hirzebruch spectral sequence (AHSS) construction of twisted K-theory. The classification of Page field strengths by twisted K-theory explains the fact that gauge invariant field strengths are classified by H-twisted cohomology, they are the twisted Chern characters [14, 15] of the corresponding twisted K-theory. They are gauge invariant because they are uniquely determined by the K-class corresponding to the physical configuration. The Page field strengths on the other hand are the integral cohomology classes that appear in the AHSS. They are not gauge invariant because the higher differentials of this sequence identify distinct cohomology classes. In other words the gauge transformations shift the Page fieldstrengths by terms that are exact under the AHSS differentials.

2.2 Classifying Solitons

In Ref. [2] Witten derived the K-theory classification of type IIB D-branes in the absence of nontorsion H flux using Sen's construction [20]. Arbitrary D-brane configurations are encoded in stacks of D9's and $\overline{\text{D9}}$'s with nontrivial gauge bundles. Up to curvature corrections the difference between the kth Chern classes of the D9 and $\overline{\text{D9}}$ gauge bundles yields the D$(9-2k)$-brane charge. Thus brane configurations are classified by pairs of gauge bundles. Tachyon condensation yields an equivalence of the pairs of gauge bundles under direct sums and so one concludes that D-branes are classified by the K-theory of the 10-dimensional spacetime. Hořava extended this construction to type IIA using non-BPS D9-branes in Ref. [21].

p-brane charges may be defined by Gauss' Law to be the integral of the corresponding $(8-p)$-form Page field strength F_{8-p} over a linking $(8-p)$-cycle. Thus if Page field strengths are classified by twisted K-theory, one may expect branes to be classified by some kind of twisted K-theory as well. The fields are integrated over a cycle that links the cycle

[5] A term in this operator was calculated in Ref. [11] generalizing an example in Ref. [5]. But it is not known whether there are further contributions.

on which the brane is wrapped, which suggests that branes are classified by some kind of dual K-theory.

In Ref. [5] MMS have argued that this dual K-theory should be the K-homology of a 9-dimensional timeslice. Their answer differs from Witten's answer, but in fact they were answering a different question. MMS considered spacetimes of the form $M^9 \times \mathbb{R}$, where the $\mathbb{R}$ is intuitively considered to be the time direction but the signature plays no role. They attempted to classify the conserved charges, that is the configurations on a 9-dimensional timeslice quotiented by the cobordisms that are dynamical processes relating configurations at two different times. These dynamical processes are the MMS instantons that will be described below. The fact that D-branes should be classified by equivalence classes of subsets and homology and not cohomology is a consequence of the construction. One begins with an arbitrary wrapping of D-branes, which defines a homology class represented by the wrapped cycle. Then one eliminates the anomalous wrappings and identifies those related by dynamical processes, leaving a quotient of a subset of homology.

On the other hand Witten classified 10-dimensional D-brane configurations, identifying those related by tachyon condensation. However, the 10-dimensions include not only the spatial directions but also the time direction. Thus while tachyon condensation is a dynamical process, the 10-dimensional configurations that are considered to be equivalent do not transform into each other over time, as the time direction has already been used in defining the configuration. Rather these configurations are somehow equivalent in the infrared. In our case we will see that they are reductions of the same M-theory configuration as seen by probes at different energy scales.

The difference between the 9 and 10-dimensional perspectives is not quite enough to explain the fact that Witten's approach concluded that D-branes are classified by a refinement of cohomology while MMS found a refinement of homology. To see this, we may consider an example with only 3 relevant dimensions and no H flux, and so that the K-theory is just ordinary integral cohomology and the K-homology is just integral homology. For example, we may consider type IIB string theory on $S^3/\mathbb{Z}_3 \times \mathbb{R}^7$ where $S^3/\mathbb{Z}_3$ is the circle bundle over the 2-sphere with Chern class equal to 3. One-branes owe their stability to

$$H_1(S^3/\mathbb{Z}_3) = H^2(S^3/\mathbb{Z}_3) = \mathbb{Z}_3. \tag{2.4}$$

In the MMS picture D1-branes are classified by the cycles that they wrap, and so by $H_1(S^3/Z_3) = \mathbb{Z}_3$. Meanwhile in Witten's picture one may ignore the six trivial time directions and consider a single D3 and $\overline{\text{D3}}$ wrapped on the lens space $S^3/\mathbb{Z}_3$. These each admit $U(1)$ gauge fields, which are characterized by Chern classes in $H^2(S^3/\mathbb{Z}_3) = \mathbb{Z}_3$. The D1-charge is just the difference between these Chern classes, and so is $\mathbb{Z}_3$-valued. Thus the two classifications appear to agree in this example. It seems plausible that in general the Poincare duality that relates K-theory and K-homology implies that these two classification schemes agree. An extension of this Poincare duality to twisted K-theory may reveal a similar agreement, although Witten's argument did not apply to the general twisted case [18].

2.3 MMS Instantons and the Conifold

If a D-brane wraps a *spin* cycle then it supports a $U(1)$ gauge field with field strength F. F is quantized but is subject to gauge transformations, some of which are large, that mix it with the NS 2-form B. The gauge invariant quantity is $B + F$. Gauge invariance implies that in particular the integral of $B+F$ over any two-cycle Σ must be single-valued, although unlike F, B is not quantized and its integral is not invariant under deformations of the 2-cycle.

As B is not gauge invariant, the integral of B over Σ is not well-defined. However if the brane's worldvolume contains a 3-manifold M^3 bounded by Σ one may define the integral by Stokes theorem. If there is a nonvanishing 3-cycle $Z^3 \supset M^3$ then Σ bounds (with opposite orientations) both M^3 and $Z^3 - M^3$ and so we may define the integral by Stokes theorem in two different ways, corresponding to two gauge choices for B. The requirement that these two definitions yield the same integral of the gauge invariant quantity $B + F$ is

$$\int_{Z^3} H + dF = \int_{M^3} H + dF + \int_{Z^3 - M^3} H + dF = (\int_\Sigma B + F) - (\int_\Sigma B + F) = 0 \quad (2.5)$$

where the minus sign came from the opposite orientations of the boundaries. In the quantum theory both the integrals of H and dF are quantized, and so (2.5) is the condition that the integral cohomology class $H + dF$ vanishes.

Freed and Witten [19] have generalized this result to the case in which a D$(p+2)$-brane wraps a manifold that is not necessarily either *spin* or *spin*^{c}. In this case there is a $\mathbb{Z}_2$ torsion correction and the condition in integral cohomology becomes

$$W_3 + H + dF = 0 \quad (2.6)$$

where W_3 is the third Stiefel-Whitney of the normal bundle of Z^3. While there is not a real gauge bundle away from the *spin* case, dF may still be defined as the Poincare dual to the endpoint of a Dp-brane in the D$(p+2)$-brane, which is the generalization of a world-volume magnetic monopole to the case in which the gauge "bundle" transition functions do not satisfy the triple overlap condition. Thus if a D$(p+2)$-brane wraps a 3-cycle with nonvanishing $W_3 + H$ then Dp-branes must end on that D-brane, and the intersection of their endpoints with the 3-cycle must be Poincare dual to $W_3 + H$. These configurations generalize the baryons of Ref. [4].

MMS [5], reducing a more cumbersome construction by Diaconescu, Moore and Witten [10], realized that the baryon is a vertex that does not conserve Dp-brane charge. In particular imagine that there is a 3-cycle Z^3 such that the third Stiefel-Whitney class of its normal bundle plus the H flux is some cohomology class[6] k

$$W_3 + H = k. \quad (2.7)$$

Now we may scatter n Dp-branes and watch for Dp-number violation. For ease of description we will consider the case in which the cycle is a 3-sphere, but the general case

[6]We will treat k as a number, but in fact the contribution from W_3 is $\mathbb{Z}_2$-torsion. However in the cases of interest this clarification will not be important.

is no more difficult, one need only use a Morse function on Z^3 to create cross-sections, as is done in Ref. [8]. The 3-sphere, however, has the extra advantage that this story may be quantitatively studied in the case of D0-branes in the SU(2) WZW model as we review in Sec. 3.

We will consider Dp-branes that initially intersect the 3-sphere transversely. A single Dp-brane, far away from the others, is then locally just a point on the S^3. The positions of two Dp-branes are encoded in a two by two matrix and as they approach each other the potential terms that suppress the off-diagonal elements become small. Once these potential terms are small enough that they do not dominate over the contribution of the H flux, the two Dp-branes blow up into a dielectric (anti-) D$(p+2)$-brane whose intersection with the 3-sphere is a surface. For simplicity we may consider $SO(3)$-symmetric configurations and so the D$(p+2)$-brane wraps a 2-sphere at constant latitude with respect to the south pole, as such 2-spheres are invariant under the $SO(3)$ isometry. This D$(p+2)$-brane carries no net D$(p+2)$-brane charge, as antipodal points on the S^2 carry opposite charges. However its worldvolume $U(1)$ gauge bundle has a Chern class of two measured in the gauge in which the integral of B over the 2-sphere is the integral of H in the 3-ball to its south. This implies that it still carries the original Dp-brane charge.

If instead we choose the gauge in which the integral of B over the 2-sphere equals minus the integral of H over the 3-disk to its north than we have performed the gauge transformation

$$\int_{S^2} B \longrightarrow \int_{S^2} B' = -\left(\int_{S^3} H - \int_{S^2} B\right) = -k + \int_{S^2} B. \tag{2.8}$$

The gauge invariance of $B+F$ then yields the transformation of the Dp-brane charge, which is minus (because of the orientation that we have chosen for the D$(p+2)$-brane) the integral of F over the 2-cycle

$$\begin{aligned} Q_{Dp} &= -\int_{S^2} F = -\int_{S^2} (B+F) + \int_{S^2} B \\ &\longrightarrow -\int_{S^2} (B+F) + \int_{S^2} B' = -\int_{S^2} (B+F) - k + \int_{S^2} B = Q_{Dp} - k \end{aligned} \tag{2.9}$$

and so we already see that Dp-brane charge is only conserved modulo k. We will see this more directly momentarily. Note however that, as will be the case for the Klebanov-Strassler geometry, if this H flux is sourced by a stack of NS5-branes then this same process will lead to a Dp-charge on the NS5-brane worldvolume that precisely compensates for the Dp-charge nonconservation above.

As more Dp-branes are added, the latitude of this two-sphere (or the value of the Morse function in the case of a general 3-cycle) increases. For a generic number n of Dp-branes the latitude depends on specific details of the configuration. An interesting latitude to consider is 90 degrees north. That is to say that we add branes until our 2-cycle sweeps out the entire 3-cycle and we are again left with a stack of Dp-branes. Like the initial configuration, there is a natural choice of gauge for B, the gauge in which B is finite and so its integral over a point vanishes. However in this case that choice of gauge corresponds to integrating H

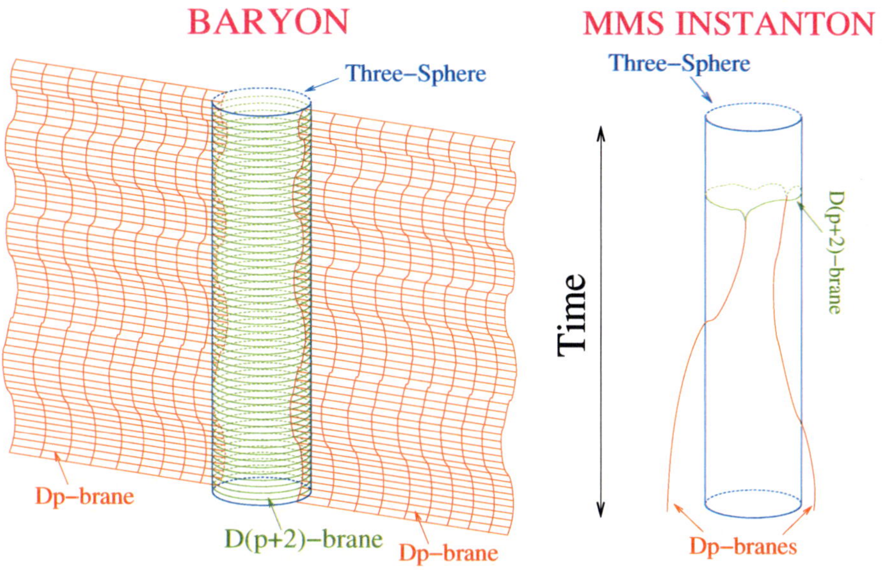

Figure 1: The baryon and MMS Instanton of type II string theory on $\mathbb{R}^7 \times S^3$ with $\int_{S^3} H = 2$. The 2 units of H-flux imply that 2 Dp-branes must end on every D(p+2)-brane wrapping the S^3. The MMS instanton violates Dp-brane charge conservation by 2 units. Charges are time-independent in the baryon configuration, and in fact the baryon number itself is conserved as it represents a nontrivial homology class $[1] \in \mathrm{H}_3(S^3)$, albeit a class that does not lift to K-homology.

over the region to the north of the 2-cycle, which has no volume and so gives zero as seen in Eq. (2.9). This is not the same gauge choice in which the integral of B over the south pole was zero, it differs by k. Thus the integral of F also differs by k, and k units of Dp-brane charge have disappeared. MMS found the same answer using the baryon construction above: a D$(p+2)$-brane has swept out a 3-cycle with k units of $H + W_3$ and so anomaly cancellation requires k Dp-brane insertions. The other ends of these Dp-branes are in the past (instead of at spatial infinity as in the baryon of Ref. [4]) and so this baryon, also known as a MMS instanton, consumes the k Dp-branes. Both configurations may be found in Fig. 1.

In addition to consuming Dp-branes the MMS instanton also leaves behind a residual flux. The D$(p+2)$-brane wrapped a noncontractible cycle and so, like an electron and positron that relatively wrap a circle and then annihilate, it leaves minus one unit of F_{6-p} RR field strength in the place of the Dp-branes that it has annihilated. The residual Page field strength is essential in demonstrating the gauge-invariance of G_{8-p}. Recall that the original Dp-brane charge can be measured by Gauss' Law by integrating F_{8-p} over a linking $(8-p)$-cycle M^{8-p}. By Stoke's theorem, the Dp-brane charge is equivalently measured by a violation of the Bianchi identity $dF_{8-p} = Q_{Dp}$. Thus as the Dp-branes are consumed, the integral of dF_{8-p} over a transverse cycle changes, where the transverse cycle wraps the

3-cycle because the Dp-branes were transverse to the 3-cycle. This is consistent with the fact that Page charges are quantized but not conserved.

On the other hand the gauge invariant field strength G_{8-p} needs to be conserved. In particular, the integral of dG_{8-p} over any two surfaces bounded by M^{8-p} must agree. We will compare this integral on a surface that intersects the Dp-branes with the integral on a surface at a later time when the Dp-branes have already decayed. These two calculations of $\int_{M^{8-p}} G_{8-p}$ agree thanks to the new flux

$$\Delta(dG_{8-p}) = \Delta(dF_{8-p}) - \Delta(H \wedge G_{6-p}) = -k - H\Delta(G_{6-p}) = -k - k\cdot(-1) = 0 \quad (2.10)$$

where we have used the fact that H does not change during this process. Equivalently the gauge invariance of G_{8-p} implies the vanishing of the integral of dG_{8-p} over any closed surface. Eq. (2.10) is the evaluation of the integral over a surface that links the instanton.

Depending on the specifics of the geometry the MMS instanton may also create a non-quantized field strength along the noncompact directions. While these cannot be determined from the topology alone, they play a role in the gauge theory as they determine the super-potential.

At the end of this process we have a stack of Dp-branes, and so the same process may repeat again, and with each iteration the stack will lose k units of Dp-brane charge and also a unit of F_{6-p} flux. In the SU(2) WZW model this process recurs until the total brane charge is less than k units. We claim, generalizing the claim of Ref. [13], that an example of this series of processes in which Dp-brane charge is lost and F_{6-p} flux is gained is S-dual to the KPV description of the Klebanov-Strassler cascade. At each step of the cascade k D3-branes in a G_3 background blow up into a dielectric NS5-brane which sweeps out the 3-cycle of the deformed conifold and disappears, leaving behind a unit of H flux along the 3-cycle and nonquantized H flux in the other directions that determines the superpotential. The G_3 background is created by 5-branes wrapping a contractible trivial 2-cycle. Because this 2-cycle is contractible the G_3 flux that it creates does not represent a cohomology class of the conifold, rather the G_3 flux felt by a D3-brane depends on the cycle it wraps and also its linking number with the D5-branes. These linking numbers, in turn, are determined by the choice of the vacuum in the low energy effective gauge theory. The baryonic vacua correspond to D3-branes that always link all m of the D5-branes and so the rank of the gauge groups falls by m at each step in the cascade.

2.4 More Baryons

One may, as in Ref. [4], consider wrapping other things over cycles in the conifold to look for varieties of baryons. For example one may wrap a D5-brane around the $S^3 \times S^2$, which supports n units of G_5 flux. The $C_4 \wedge B$ term in the D5-brane worldvolume action then implies that n fundamental strings must end on the D5. If these strings have a finite energy and so a finite length then the other ends must be on the branes in which the $SU(n) \times SU(n+m)$ gauge theory lives, and so we have found a vortex of n particles in the gauge theory. Numerology suggests that in the worldvolume gauge theory this is a kind of baryon vortex coupling n external charges that each transform in the fundamental representation of $SU(n)$. In the Klebanov-Strassler geometry the conifold is deformed, which means that a sufficiently low-energy probe notices that the 2-sphere is contractible and so this D5-brane carries no net charge. Thus this description of the baryon breaks down at low energies, as may be expected from the fact that the $SU(n)$ gauge group does not survive the cascade.

One may also wrap a D3-brane around the 3-sphere. The 3-sphere supports m units of G_3 flux and so m fundamental strings must end on the D3-brane. In addition there is an energy-dependent amount of H flux on the 3-sphere, which suggests that some number of D-string insertions may be required as well. In the gauge theory there then appears to be a vortex coupling m external charges, with possibly an energy-scale dependent number of magnetic monopoles. Such a vortex may be constructed by contracting the epsilon tensors of the $SU(n)$ and the $SU(n+m)$ gauge groups, in other words it is a kind of bound state of an $SU(n+m)$ baryon and an $SU(n)$ antibaryon. These baryons have appeared in Ref. [7].

So what about the $SU(n+m)$ baryon? It must somehow be the sum of the two baryons just constructed, but on the other hand it would be surprising if it looked very different from the $SU(n)$ baryon. To superimpose the above two baryons, one need only note that the D3-brane charge of the 5-brane is just the integral of the U(1) field strength over the 2-sphere. The normal bundle of the 2-sphere is not *spin*, and so the integral of the field strength obeys the shifted quantization condition

$$\int_{S^2} F \in \mathbb{Z} + \frac{1}{2}. \tag{2.11}$$

This shifted quantization condition is responsible, for example, for the quantum Hall effect that leads the fermions of Ref. [22] to exhibit bosonic statistics. Thus the D5-branes always carry half-integral D3-brane charge. The $SU(n+m)$ baryon is a D5-brane whose worldvolume U(1) gauge field has a Chern class one higher than the $SU(n)$ baryon, while the $SU(n-m)$ baryon has a Chern class that is one lower. All of these baryons exist at every energy scale, as a bundle exists with every half-integral Chern class. In the gauge theory, these are all just boundstates of the $SU(n)$ baryon and some number of $SU(m)$ baryons. One cannot try to calculate the flux on a D5-brane that corresponds to a specific baryon, because F is subject to large gauge transformations. $F+B$ is gauge invariant, and a natural choice for B shifts by a unit each time an NS5-brane passes by, that is each step in the cascade. And so at each step in the cascade the amount of flux corresponding to each choice of baryon shifts by one, which compensates for the shift in the gauge group.

Notice that the mysterious D-strings ending on the $SU(m)$ baryons must end on the NS5-branes, because of the G_5 flux created by the D3-branes and the $C_4 \wedge C_2$ coupling on the NS5-brane worldvolume. Thus at each step in the cascade, when an NS5-brane passes through, one of these D-strings shrinks to zero length and then never re-emerges after the NS5-brane has passed. This is just the Hanany-Witten transition [23]. Then the number of monopoles coupled by the vortex at a given energy scale only depends on the difference between the rank of the vortex and the rank of the gauge group at that energy scale and not on any arbitrary initial choice of energy scale. Up to a constant shift, there is one monopole for each m-index epsilon tensor contracted with the original $n+m$-index tensor. This may be seen directly from the

$$S_{D5} \supset \int C_2 \wedge B \wedge (B+F) \tag{2.12}$$

term in the D5-brane worldvolume action. The integral of one $B+F$ over the 2-sphere counts, up to a constant and typically nonintegral shift, the number of m-epsilon tensors contracted with the epsilon tensor of rank equal to the rank of one of the gauge groups at the energy scale at which B is measured. The $C_2 \wedge B$ term then, as usual, implies that for each unit of H flux there is a D-string ending on each of these D3-branes, which agrees with the fact that for each unit of D3-brane charge there is a D-string ending on each NS5-brane.

3 Dielectric NS5-Branes from the $SU(2)$ WZW Model

The $SU(2)$ WZW model at level $m-2$ describes string theory on the $SU(2)$ group manifold with m units if H flux. Using boundary conformal field theory a number of authors have shown (see for example [24] for a description of the final states) that if a stack of n D0-branes is placed on the 3-sphere then the H flux causes the D0-branes to fatten into dielectric D2-branes that wrap a 2-sphere at a constant latitude. The latitude is only known in the semiclassical (large m) regime [25], where it has been found to be $n\pi/m$ using the Born-Infeld action on the D2-brane. However the boundary CFT approach holds for any value of m and demonstrates that as the number of D0-branes increases, the latitude increases and upon adding the $m-1$st brane the spherical D2-brane collapses to an anti-D0-brane at the north pole. If we add one more D0-brane, for a total of m, it annihilates this anti-D0 and no branes remain. If instead we started with $m+k$ D0-branes we would then find that m of them sweep out the 3-sphere and then self-destruct as above, leaving k, which themselves form a spherical D2-brane. If $k>m$ then the new D2-brane sweeps out the sphere again and so on. Thus we find a cascade, the D0-branes blow up into spherical D2-branes which sweep out the sphere repeatedly until the D0-brane charge is less than m, at which point a single spherical $D2$-brane remains.

The $SU(2)$ WZW model alone does not describe a critical string theory, however it may be a component of a critical string theory. For example the near-horizon limit of a stack of m NS5-branes and F-strings wrapped on a 4-torus yields string theory on $AdS^3 \times S^3 \times T^4$ with m units of H flux supported on the S^3. The S^3 component is then described by a level $m-2$ $SU(2)$ WZW model, even away from large m and without taking the near-horizon limit. One can then watch the cascade. For example, one may place a stack of n Dp-branes at a point on the S^3 with any extra directions extended along the T^4 and watch the system relax. The Dp-branes will blow up into dielectric D$(p+2)$-branes and sweep out the 3-sphere as many times as they can, each time losing m units of Dp-brane charge. The worldvolume $SU(n)$ theory on these Dp-branes will then cascade, although the extra Dp-charge of the NS5-branes means that we need to be careful when calculating the effect on the gauge group.

Instead of watching the cascade occur over time, one may watch it occur as the radius varies. That is one may consider a time-independent configuration in which the Dp-branes extend from the horizon (or from infinity is we do not take large m and the near-horizon limit). One may then take cross-sections at different radii and compare the Dp-brane configurations. To do this one must specify some kind of boundary condition, we will choose a condition in which at some fixed radii the cross-section contains n Dp-branes that are on top of each other and not blown-up into dielectric D$(p+2)$-branes. The energy of a configuration at a smaller radius is then just the energy from the WZW-model plus a correction from the derivative of the configuration with respect to the radius, which is negligible so long as the configuration relaxes slowly enough with respect to the radius. Thus as we move in closer to the origin, we see the Dp-branes blowing up and sweeping out the 3-sphere again and again until finally they carry less than m units of Dp-brane charge, at which point they form a spherical D$(p+2)$-brane that roughly sweeps out a cone as it dives into the origin. From the point of view of the worldvolume theory of the NS5-brane, cross-sections of smaller radii correspond to lower energies. Thus decreasing cross-sections correspond to flowing into the IR of the worldvolume theory, and so like the Klebanov-Strassler cascade, this cascade proceeds as one flows into the infrared. In the large m or n limit this is the familiar fact from the AdS/CFT correspondence that the IR corresponds to small radii.

Different values of p are related by T-duality on the 4-torus. If we consider $p = 3$ and then S-dualize we find the desired D3-branes in a background G_3 flux expanding into dielectric NS5-branes and sweeping out 3-spheres. This is not the same as the Klebanov-Strassler cascade because the spacetime contains a cone over S^3 and not over the 5-manifold $T^{1,1}$, however the extra 2-sphere in the conifold appears not to qualitatively interfere with the dielectric effect on the S^3. In the conifold case we generally adopt the perspective in which we watch the system relax in time rather than using boundary conditions to orchestrate the evolution as a function of radius that may be interpreted as an RG-flow. However in the T-dual MQCD perspective, when we consider a finite string coupling by including the M-theory circle or by allowing the branes to bend, we see that this RG-flow is just a radial-dependence of a supergravity solution. Thus we expect that, by T-duality, if we include finite-coupling effects by lifting the conifold to F-theory we will see the entire cascade at once. In particular the D3-branes lift to F-theory 5-branes[7] whose embedding depends on the torus coordinates such that going about one of the cycles of the torus (the IIA cycle) the configuration shrinks while the branes sweep out the 3-sphere.

4 The Cascade from MQCD Brane Cartoons

The Klebanov-Strassler realization of the cascade using D3 and D5-branes on the conifold is T-dual to an equivalent brane cartoon in IIA. Such brane cartoons do not require large N as one may learn about the gauge theory without using the dual bulk supergravity, semiclassical information is contained in the parameters of the brane cartoon. Quantum mechanical

[7]While in the IIB description the dielectric effect appears to be a dramatic transformation of D3-branes into 5-branes, in this particular F-theory description like in the M-theory description one may expect a single brane.

features of the configuration, such as the RG flow and the theta angle, are encoded in the cartoon's M-theory lift.

The reduction of an M-theory configuration to a IIA configuration does not necessarily exist and in general is not unique. We will see that for configurations whose low energy effective theories describe the baryonic root of superQCD the reduction is determined by the energy scale considered. Considering progressively lower energy scales one obtains different IIA brane cartoons that correspond to different stages in the cascade. Each of these cartoons may then be T-dualized to obtain a corresponding configuration of branes and fluxes in the Klebanov-Strassler geometry, and so one may watch the cascade progress in IIB. In the KPV proposal this effect is replicated by introducing anti D3-branes.

The description of the IIB cascade that we find will be a cousin of the MMS instanton. Rather than the D3-branes expanding into dielectric NS5-branes at fixed radii, which would break the symmetries of the conifold and therefore perhaps some internal symmetries of the gauge theory, we will see a symmetric RG-flow trajectory. The RG-flow trajectory produced by MQCD and the symmetric reduction to IIA that we will choose is T-dual to a conifold geometry in which n D5-branes wrapping the trivial 2-cycle contract. In contracting they sweep out H flux and so the integral of the NS 2-form B over the two-cycle monotonically shrinks. Each time that B shrinks by one unit, this is gauge equivalent to the integral of the field strength F of the U(1) gauge field on each D5-brane decreasing by one. Making this gauge transformation to keep B in the fundamental domain $B \in [0, 2\pi)$ we see that each D5-brane loses one unit of D3-brane charge each time that a unit of H flux is swept out. The result is that m units of D3-brane charge are lost at each step. The cascade stops when there is no more H flux to link. The baryonic root corresponds to the configuration in which each D5-brane links the same maximal amount of H flux, but as is clear in the M-theory picture there are many other choices of vacuum available.

4.1 The $N = 2$ Cascade in MQCD

Consider "weakly coupled" type IIA string theory on $\mathbb{R}^9 \times S^1$ where the compact direction is $x^6 \sim x^6 + 1$. Extend two NS5-branes along x^{0-5} both at $x^{7-9} = 0$ but at distinct x^6 coordinates $x^6 = a - 1/2$ and $x^6 = 1/2 - a$. Let n D4-branes wrap the circle and also extend along x^{0-3} while another m are parallel but only extend from $x^6 = a - 1/2$ to $x^6 = 1/2 - a$. For now we will place all of the D4-branes at $x^{4,5,7-9} = 0$.

Open strings may connect various D4-branes and in a limit in which the string coupling and the circle are small these decouple from other excitations and describe a four-dimensional $N = 2$, $SU(n) \times SU(n+m) \times U(1)$ gauge theory [26]. In fact even the $U(1)$, which describes center of mass motions of the system of D4-branes, decouples and we will often omit it. Fundamental strings connecting D4-branes without crossing NS5-branes yield the vectormultiplets. In addition there are 2 bifundamental $SU(n) \times SU(n+m)$ hypermultiplets described by F-string-like objects (their M-theory lifts are discs and so do not wrap the entire M-theory circle) connecting D4-branes on opposite sides of an NS5.

Quantum features of the gauge theory arise from g_s effects in the brane cartoon, such as the fact that the D4-branes pull on the NS5-brane. To include these effects we will lift our

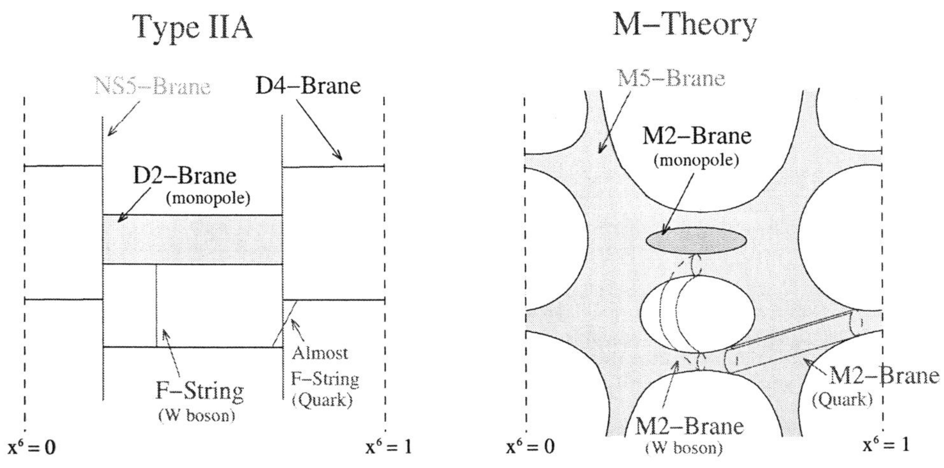

Figure 2: The type IIA and M-theory realizations of a $SU(2) \times SU(3)$ $N = 2$ supersymmetric gauge theory, with 2 bifundamental hypermultiplets. In IIA the theory and vacuum are described by D4 and NS5-branes, while the degrees of freedom are F-strings and D2-branes. In the M-theory lift the theory and vacuum are described by an M5-brane while the degrees of freedom are M2-branes. Here all D4-branes end on NS5-branes, but in general a D4-brane may also wrap the circle x^6.

configuration to M-theory, and so the D4-branes and NS5-branes will be replaced by M5-branes. This lift is not unique as two D4-brane segments separated by an NS5-brane may either connect to each other or else one connects to the NS5-brane while the other connects to another adjacent D4-brane. These choices correspond to the choices of different vacua in the gauge theory. Here we will consider the two simplest choices, the $r = 0$ nonbaryonic root in which a minimal number of D4-branes are connected across NS5-branes and also the case with the maximal number of connections, which leads to a cascade of Seiberg dualities in which the baryonic root is selected every time. In the $N = 2$ theories considered here there are continuous moduli spaces of vacua, but these vacua are interesting as they will be among a finite collection of vacua that survive the superpotential perturbations that we will consider later.

We will describe the M-theory lifts in terms of the complex coordinates

$$t = \exp(x^6 + \frac{ix^{10}}{R}), \qquad v = x^4 + ix^5 \tag{4.1}$$

where x^{10} is the M-theory circle, which has radius R. As the x^6 coordinate is periodic, different values of t will correspond to the same physical point. All of the M5-branes extend along the gauge theory directions x^{0-3}, and so configurations will be determined by the additional Riemann surface that they wrap in the $t - v$ plane.

The $r = 0$ nonbaryonic root vacuum consists of $n + 1$ M5-branes. n of these are flat and wrap the torus $v = 0$, while the last wraps the Riemann surface

$$t^2 + (v^m - 2(-\Lambda)^m)t + \Lambda^{2m} = 0, \qquad |t| < e^{1/2} \tag{4.2}$$

where the bound on t will be explained below. The maximal cascade configuration is quite similar. Again one M5-brane wraps the logarithm of the Seiberg-Witten curve

$$t^2 + (v^m - 2(-\Lambda)^m)t + \Lambda^{2m} = 0, \qquad |t| < \exp(\frac{\lfloor \frac{n}{m} \rfloor + 1}{2}) \tag{4.3}$$

where $\lfloor x \rfloor$ is the greatest integer less than or equal to x. In addition

$$k = n - \lfloor \frac{n}{m} \rfloor m \tag{4.4}$$

M5-branes wrap the torus $v = 0$. We will argue that if a superpotential polynomial in the adjoint chiral multiplet is added then the cascade results in an $N = 4$ $U(k)$ gauge theory.

The couplings of the gauge theories are inversely proportional to the square of the distance between the NS5-branes in x^6, and so small couplings correspond to $\Lambda << 1$. In this limit it is possible to reverse the above process and dimensionally reduce each M-theory configuration to a IIA brane cartoon at $g_s = 0$. One may then read the gauge couplings, the ranks of the gauge groups and other semiclassical details of the theory off of the IIA brane cartoon. The complication is that the logarithmic shape of the M5-brane means that the x^6 positions of the resulting NS5-branes are ill-defined. However given a distance scale E, which corresponds to an energy scale proportional to E because the probe M2-branes have a fixed tension, one may define a reduction to IIA as follows [27]. Consider the points on the M5-brane that are at $|v| = E$, these consist roughly of two circles in the above limit $\Lambda << 1$, where we ignore the gauge theory directions. These circles are at roughly fixed values of x^6. We place the two NS5-branes at these two values of x^6 and consider the M5-brane at smaller values of v to be D4-branes connecting those two NS5-branes at $v = 0$. The reduction of the flat toroidal components of M5-branes is well defined, they yield D4-branes wrapping the entire x^6 circle.

As seen in Ref. [26] the distance between the NS5-branes, which determines the gauge coupling, depends logarithmically on the energy considered and so the RG flow is reproduced. However when these two circles pass through each other the number of resulting D4-branes changes so that Seiberg duality is also reproduced [27, 28]. The bounds on t in Eqs. (4.2,4.3) can now be seen to reflect the fact that given the gauge theory at some energy scale one can only determine (up to corrections that are exponentially suppressed in the allowed region) the part of the curve at lower values of t, which correspond to flowing the theory into the IR. An attempt to continue into the UV is obstructed by the fact that, as one of the two gauge groups is strongly coupled in the UV, one needs to determine a UV completion. This UV completion is not unique, but rather depends on the connections of D4-branes to NS5-branes at higher energy scales than the one in which we specified our

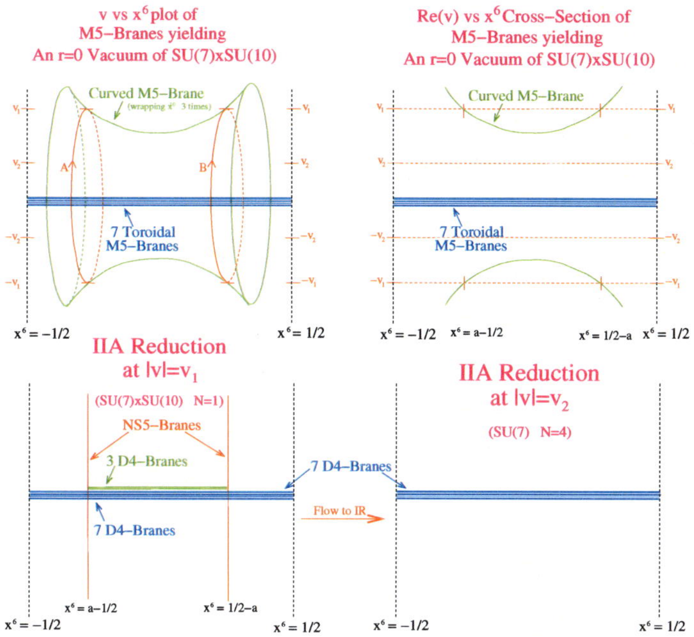

Figure 3: The type IIA and M-theory realizations of the $SU(7) \times SU(10) \longrightarrow SU(7)$ cascade. This is not a Klebanov-Strassler type cascade, which would end with only the decoupled $U(1)$ as seen in Fig. 4. On the upper-left v and x^6 are represented, while the upper-right is a projection of the same configuration onto Re(v) and x^6. The reduction to IIA is performed at two energy scales, v_1 and v_2. The intersection of the M5 with $|v| = v_1$ consists of the loops A and B, which have winding numbers of 3 and -3 about the M-theory circle. Thus the reduction at that energy scale yields 2 NS5-branes, one connected to 3 D4's on the right and the other to 3 D4's on the left. The M5-brane does not intersect $|v| = v_2$, and so a reduction at that energy scale does not result in any NS5-branes. In every cartoon we see the 7 flat branes that support the $SU(7)$ $N = 4$ theory at the bottom of the cascade.

initial conditions. That is to say that two different vacua in two different UV theories may flow to our starting point.

In general any UV completion involves an infinite number of cascades. An infinite number of cascades would be avoided if, for example, m D4-branes connect to the NS5-branes far away on the v-plane, corresponding to m extra vector multiplets with large adjoint scalar VEVs that in the UV combine with our $SU(n) \times SU(n+m)$ to yield the conformal theory

$$SU(n) \times SU(n+m) \times U(1)^m \xrightarrow{UV} SU(n+m) \times SU(n+m). \tag{4.5}$$

However such a completion could be apparent in the IR, as the m extra vector multiplets would be massless although weakly coupled to the rest of the theory. If some of the UV theories are in nonbaryonic root vacua then the number of extra D4-branes required to stop the cascade will generically be different from m.

We may now apply the above dimensional reduction procedure to understand the RG flows of the vacua (4.2) and (4.3). In the first case, which corresponds to the $r = 0$ nonbaryonic root of the $SU(n+m)$ $N = 2$ gauge theory with $2n$ flavors of hypermultiplets, the circles corresponding to any v in the allowed domain never cross, and so at every energy down to the strong coupling scale E (defined to be the shortest distance between the curved and flat M5-branes) we find two NS5-branes connected by m D4-branes and another n D4-branes that wrap the entire circle. In IIA we may equivalently say that n D4-branes run between the two NS5-branes in one direction while $n+m$ run in the other direction, and so we have the original $SU(n) \times SU(n+m)$ theory. At the energy scale E the $SU(n+m)$ is strongly coupled as the corresponding NS5-branes coincide.

It is tempting to then claim that at smaller energies, smaller than the minimal radii of the circles, there are no NS5-branes and so we are left with the $SU(n)$ theory on the n D4-branes wrapping the circle. However the circles cease to be approximately circular and in fact extend to arbitrarily small v ($E = 0$ in the $N = 2$ case) at these energies and so this argument fails. The curve becomes singular as the cycles that bound massless dyons degenerate. Yet if, as seen in Fig. 3, we turn on a superpotential for the adjoint chiral multiplet, which gives VEVs to these massless dyons[8], then the circles no longer extend to arbitrarily small v and one may conclude that the $SU(n+m)$ gauge symmetry, after it becomes strongly coupled, is gapped (is Higgsed by the dyon VEVs)

$$SU(n) \times SU(n+m) \stackrel{r=0}{\longrightarrow} SU(n). \tag{4.6}$$

This is as one expects for the $r = 0$ vacuum, after all r is the rank of the remaining gauge group in the $SU(n+m)$. This $SU(n)$ gauge theory lives on the worldvolume of the D4-branes that wrap the torus, and the light degrees of freedom do not couple to the far away 5-branes. Thus the supersymmetry is enhanced to that of a stack of flat, parallel branes, $N = 4$ [7]. In fact the final gauge symmetry is $U(n)$, we have been consistently ignoring the extra $U(1)$ that describes the center of mass of the D4-branes.

4.2 The $N = 1$ Cascade

We will now, as described above, introduce a nondegenerate superpotential $W(\Phi)$ for the adjoint chiral multiplet Φ, thus giving VEVs to the massless dyons, resolving the singularities and allowing reductions to type IIA. By nondegeneracy we mean not only that the extrema are distinct, but that the distances between the extrema are larger than the energy scale considered. These two conditions ensure that it suffices to expand the superpotential to second order, and so the physics is identical to that of the $N = 2$ theory with supersymmetry softly broken by a (possibly infinite) mass term for the adjoint chiral multiplet.

[8]At the baryonic root VEVs are given to a maximal set of these massless dyons, deforming away all of the singularities.

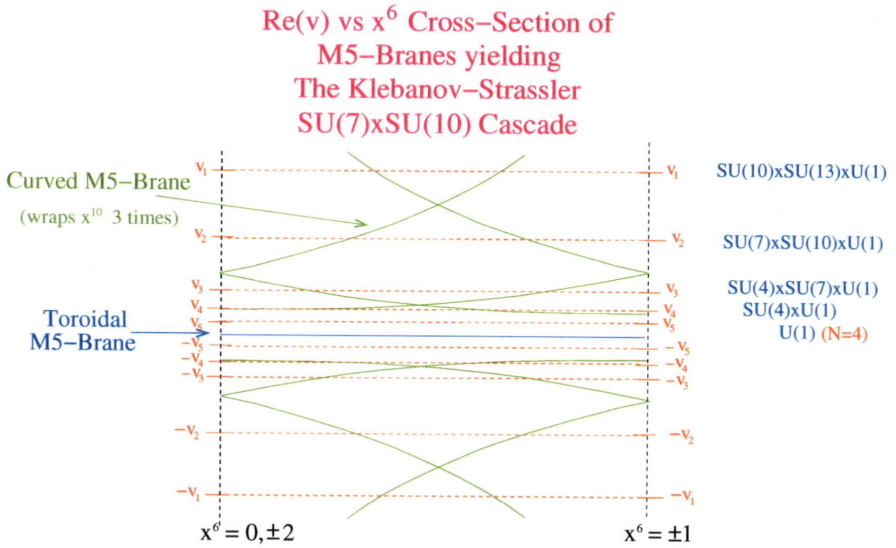

Figure 4: Here we see the M-theory configuration corresponding to the Klebanov-Strassler cascade $SU(7) \times SU(10) \longrightarrow U(1)$, where the $U(1)$ is present at every stage but generally omitted. The continuation to the higher energy scale v_1 is not unique, and we have chosen the continuation such that the UV $SU(10) \times SU(13)$ is in a baryonic root vacuum. The effective gauge group at each energy scale v_k is written next to the energy scale. When the group is $SU(a) \times SU(a+b)$ then the IIA reduction consists of two NS5-branes with a D4-branes connecting them on one side and $a+b$ on the other. Notice that this M5-brane configuration is different (only 1 flat brane) from the M5-brane configuration in Fig. 3, which describes a different vacuum and so a different cascade of the same initial $SU(7) \times SU(10)$ theory.

Introducing the new complex coordinate $w = x^8 + ix^9$ the effect of a superpotential is merely to change the x^8 and x^9 coordinates of the NS5-branes from $w = 0$ for both to

$$w = 0 \qquad \text{and} \qquad w(v) = W'(v) \tag{4.7}$$

respectively. In particular a superpotential does not affect, in our small Λ approximation, the pattern of t coordinate crossings. The superpotential does have the effect of lifting most of the moduli space, leaving adjoint scalar VEVs either at the extrema of the superpotential, at which $w(v) = 0$, or else at the values of the bare masses of the flavors to which they are locked. In the first case the analysis proceeds as above but now with a potential cascade at each critical point. The compactness of x^6 means that the second case corresponds to flat D4-branes wrapping the entire configuration but not ending on an NS5-brane. While such branes contribute to the rank of the gauge group, they do not affect the cascade as they are not connected to the participating branes. The $N = 1$ case has the advantage that all of the vacua except for the $r = 0$ vacua have some form of multi-step cascade, unlike the $N = 2$ case above in which cascades only occurred at very special points in a huge moduli space.

Klebanov and Strassler [7] have treated the case in which the NS5-branes are orthogonal. This is a limiting case of the above construction in which a mass term is included for the adjoint chiral multiplet $W(\Phi) = \mu\Phi^2$ and then the mass μ is taken to infinity.

Now we are ready to study the curve Eq. (4.3), drawn in Fig. 4, which has been seen in Ref. [7] to yield the Klebanov-Strassler cascade. This claim is easily reproduced by the above algorithm. For concreteness, we will describe a cascade that begins with a $SU(7) \times SU(10)$ gauge theory, so that

$$n = 7, \qquad m = 3, \qquad k = 1, \qquad |t| < e^{3/2}. \tag{4.8}$$

Further we consider a small adjoint scalar mass μ so that the projection onto the surface $w = 0$ is just (4.3). This choice does not affect any qualitative features of what follows, and in fact the opposite limit has already been used in Ref. [7]. At every energy scale there will be $k = 1$ D4-brane wrapping the x^6 circle, corresponding to the flat M5-brane component not connected to the Seiberg-Witten curve. Following the above algorithm, we find the positions of the other branes by looking for circles at fixed $|v|$ on the M5-brane component

$$t^2 + (v^3 + 2\Lambda^3)t + \Lambda^6 = 0 \tag{4.9}$$

which may be solved for t

$$t = -\Lambda^3 - \frac{v^3}{2} \pm \frac{\sqrt{(v^6 + 4v^3\Lambda^3)}}{2}. \tag{4.10}$$

The M5-brane curve enjoys a $x^6 \leftrightarrow -x^6$ symmetry which translates to $t \leftrightarrow 1/t$. The small Λ approximation leads one gauge group to be weakly coupled while the other undergoes Seiberg duality and may be used to eliminate Λ at $|t| > 1$. Thus we may find the NS5-brane at $|t| > 1$ and use the $t \leftrightarrow 1/t$ invariance of our configuration to find the other NS5-brane.

To zeroth order in Λ this yields $t = -v^3$ at $|t| > 1$, in accord with the fact that each NS5 is connected to 3 D4's. Thus each absolute value of v, corresponding to the energy scale $E = |v|$, intersects the M5-brane on the circle $|t| = E^3$. The NS5-branes are therefore placed at

$$x^6 = \pm 3\ln(E). \tag{4.11}$$

Note that the distance between the two NS5-branes reproduces the known RG flow of the $N = 2$ $SU(n+m)$ theory

$$\begin{aligned} \frac{1}{g_{YM}^2} \sim \Delta x^6 = 3\ln(E) - (-3\ln(E)) = \\ 6\ln(E) = (2(n+m) - 2n)\ln(E) = (2N_c - N_f)\ln(E) \end{aligned} \tag{4.12}$$

as expected from the fact that at scales greater than μ this theory is very close to the $N = 2$ theory. The NS5-brane at positive x^6 is connected to $m = 3$ D4-branes, all on its left, while that at negative x^6 is connected to 3 D4-branes on its right. These are the same 3 D4-branes, but the number of times that the D4-branes wrap the x^6-circle before connecting depends on $|v|$ because the total number of D4-branes at any cross-section of constant x^6 must be $m = 3$ times the number of M5 windings with radius less than $|v|$ plus the number of flat M5's. In the original theory, which corresponds to $1 < |x^6| = 3ln(E) < 3/2$ or equivalently $e < |t| < e^{3/2}$ (and its reciprocal) we find that the 3 D4-branes are emitted from one NS5-brane, make two full circles around x^6 and then finally continue a little farther to end on the other NS5-brane. Thus in addition to the omnipresent disconnected D4-brane, there are an extra 9 D4-branes in the region that the 3 D4's wrap three times and an extra 6 in the region wrapped only twice. We then find that one region contains $7 = 1+6$ D4-branes and the other $10 = 1 + 9$, reproducing the original $SU(7) \times SU(10)$ symmetry.

At $|t| = e$ the NS5-branes cross and so at $\sqrt{e} < |t| < e$ the 3 D4-branes only wrap the entire x^6 once. Thus the region in which there were always 7 D4-branes has shrunk, but still exists leaving an $SU(7)$. However the region wrapped 3 times, which was between the two NS5-branes that have now passed through each other, has now ceased to exist. Once the NS5-branes have passed each other a new region opens up between them. This one has again the omnipresent D4, but only 3 other D4's, which are the stack of D4's that wrap the entire circle. The remaining gauge symmetry is now $SU(4) \times SU(7)$.

At $|t| = \sqrt{e}$ the NS5-branes cross again and so at $|t| < \sqrt{e}$ the 3 D4-branes do not wrap the entire x^6 circle, but merely extend directly from one NS5-brane to the next as in the $r = 0$ configuration. Thus there are two regions, one with 4 D4's through which the stack of 3 passes, and the other with only the free D4-brane, yielding a $SU(1) \times SU(4)$ symmetry. $SU(1)$ is just a point, however in fact we have been consistently ignoring the $U(1)$ whose scalar VEV parametrizes center of mass motions of the free D4-brane, and so the symmetry is $U(1)$. Now we have returned to the $r = 0$ vacuum case. At energy scales yet lower, so that no probe is energetic enough to extend from the toroidal M5 to the curved M5's, the $SU(4)$ gauge symmetry is Higgsed by dyon VEVs and only the $U(1)$ $N = 4$ theory remains.

4.3 Generalization: Gauge Groups with More Simple Components

Adding extra NS5-branes leads to cascades with more than two gauge groups, but each iteration in the cascade still corresponds to two NS5-branes crossing and so the remaining gauge groups may be found as above. In particular when each gauge group becomes strongly coupled then at the baryonic root the new gauge group is just the Seiberg dual, while nonbaryonic roots correspond to some D4-branes being suspended between the coincident NS5-branes in the M-theory lift and so the corresponding colors do not appear in the lower energy effective theory. The term "Seiberg dual" is used loosely here, as in the case of noninfinite chiral adjoint multiplet mass (which is unavoidable in the case of an odd number of NS5-branes) the superpotential contains quartic terms that were not in Seiberg's original magnetic theory (see for example Ref. [29]).

The simplest class of examples of cascades with three gauge groups occurs when two NS5-branes are connected to the same number m of D4-branes on the same side. In particular if baryonic roots are chosen in every vacuum then two NS5 branes remain parallel while the third crosses them both in the same order repeatedly until the end. Then at each step of the cascade the rank of the largest gauge group is reduced by $3m$

$$\begin{aligned} SU(n)\times SU(n+m)\times SU(n+2m) &\longrightarrow SU(n)\times SU(n+m)\times SU(n-m)\\ &\longrightarrow SU(n)\times SU(n-2m)\times SU(n-m)\\ &\longrightarrow \ldots . \end{aligned} \tag{4.13}$$

The case $m=1$ would at fairly low energies give an $N=1$ theory with a $SU(3)\times SU(2)\times U(1)$ gauge symmetry, although with the wrong matter content.

More generally the period of the cascade is roughly the least common multiple of the ranks of the simple gauge groups times the number of simple groups. For example if we begin with $SU(n)\times SU(n+1)\times SU(n+3)$ with the $SU(n)$ group just below its UV fixed point and the $SU(n+1)$ group at its UV fixed point then the cascade consists of drops in rank of 1, 4 or 5 at a time, repeating the sequence: $5,4,5,4,5,4,5,4,5,1$. In this case there are three NS5-branes, one attached to a single D4-branes extending to the right ($+x^6$), one with two D4-branes extending to the right and one with 3 D4-branes extending to the left. These numbers $1,2$ and 3 are uniquely determined by the gauge group, as they are the differences in ranks between the three simple components. As we flow to the IR the first two NS5-branes move to the right, the second with twice the velocity of the first, while the third moves left. The rank of a gauge group drops by $4=3+1$ when the left-moving NS5, which is attached to 3 D4's crosses the right-moving NS5 that is attached to a single brane. Similarly a rank drops by $5=3+2$ when the left-moving NS5 crosses the right-mover attached to 2 D4-branes. A rank drops by a single unit when the right-mover attached to 2 D4-branes overtakes the other right-mover.

In the case of more than 3 simple components the number of D4-branes attached to a given NS5-brane is not uniquely determined by the gauge group, but is determined by the gauge group together with the bifundamental matter content. The number of D4-branes attached to a given NS5 is then the difference between the ranks of two simple components such that there is bifundamental matter charged under the two components. At each step of the cascade two NS5-branes cross, which causes the rank of a simple component of the gauge group to drop by the difference in the number of D4-branes ending on those two NS5-branes. Here D4-branes attached to opposite sides of NS5-branes are counted with opposite signs. We count only D4-branes whose corresponding adjoint scalar VEVs are at the same extremum of the superpotential. Sufficiently separated minima have cascades which may be treated separately. The treatment of nearby minima and in particular degenerate critical points of the superpotential is beyond the scope of this note.

At large N the endpoints of such transitions have been found explicitly in Ref. [30], who also used both the MQCD approach and its T-dual IIB realization.

5 T-Dualizing the MQCD Cascade

5.1 The Cascade from a Decreasing B-Field

We will now T-dualize the x^6 circle of the cascading IIA brane cartoons to find the cascade in IIB. On the IIA side there are two interpretations of the cascade. First, different steps in the cascade are different approximations of the same M-theory configuration. Later steps in the cascade correspond to the M-theory configuration as seen by less energetic M2-brane probes. The second interpretation is that early steps in the cascade correspond to initial conditions in type IIA string theory in a bizarre kind of weak-coupling limit where $g_s \to 0$ prevents the NS5-branes from bending, but at the same time the tension of the D4-branes causes them to contract over time, pulling the NS5's. As we time-evolve the system the NS5-branes are pulled together, finally annihilating to leave only the D4-branes that wrap all of x^6.

The M-theory configuration is contained in the F-theory configuration, one needs only to project away the IIB circle, and so the first interpretation of the cascade may allow one to find that various steps of the cascade in the Klebanov-Strassler geometry are just reductions of F-theory to IIB at different energy scales. In fact, it has already been seen that the conifold transition is smooth when the M-theory circle is fibered over it [31]. However the IIB configurations themselves, at small N, contain the same information (*i.e.* no RG flow information) as the type IIA configurations. That is, each frame is a semiclassical approximation of the quantum theory, where one approximates the effective theory at some energy scale. Thus as we watch the cascade in type IIB we will pass through successive IIB approximations at $g_s = 0$ each of which is obtained from the previous by relaxing the configuration. As in the case of M-theory reductions to type IIA, such approximations do not exist in general, for example the Argyres-Douglas [32] vacua admit no such reduction and correspondingly there is no decoupling of the full six-dimensional physics of the M5-brane worldvolume that allows the kind of four-dimensional description admitted by the baryonic root vacua.

The IIA brane cartoons are independent of the first four coordinates, in which the gauge theory lives. Therefore we need to T-dualize the remaining 6-dimensional configuration. As we are only interested in the ranks of the remaining gauge groups and the implications for the S-dual of the twisted K-theory classification, and not in a quantitative analysis of the condensates, we will only identify the topology of the dual configuration.

We will T-dualize our configuration by dividing it into four-dimensional slices $\mathbb{R}^3 \times S^1$ and then dualizing each slice using the following three facts. First, each NS5-brane that is not parallel to the S^1 is T-dual to a KK-monopole with respect to the T-dual circle, and so in particular k NS5-branes are T-dual to a A_{k-1} singularity. If the NS5-branes are at different points in the $\mathbb{R}^3$ then the singularity is blown up by topological 2-spheres that semiclassically are long cylinders with the ends pinched and so the area of a 2-sphere is proportional to the length of a tube, which is the distance between the NS5-branes in the $\mathbb{R}^3$. The distance between the NS5-branes along the x^6 circle yields a B-field on the blown-up sphere, even when the area of the sphere vanishes because the NS5's are coincident along the other three directions. Second, each D4-brane that completely wraps the circle and does not extend in any of the other three directions is T-dual to a D3-brane at a point in the T-dual four-dimensions. Finally a D4-brane extended along the circle that ends on two NS5-branes is T-dual to a D5-brane that wraps the corresponding blown-up two-sphere[9]. The normal bundle of this two-sphere has a nontrivial second Stiefel-Whitney class which leads to a shifted quantization condition for the worldvolume $U(1)$ gauge field, this is shown to be responsible for the quantum Hall effect in the worldvolume theories in Ref. [22]. The shifted quantization condition implies that the D5-branes carry half-integral D3-brane charge

$$Q_{D3} = \int_{S^2} F \in \mathbb{Z} + \frac{1}{2}. \tag{5.1}$$

A simple choice of 4-dimensional slices consists of one slice for each point v. Thus in the case of a gauge group $SU(n) \times SU(n+m)$ each slice consist of two NS5-branes separated by $W'(v)$ in the $\mathbb{R}^3$ and along the circle by the inverse coupling squared of one of the simple components. The T-dual of that slice then consists of a resolved A_1 singularity where the blown-up circle has an area that is roughly $W'(v)$ and a B-field proportional to an inverse coupling squared.

In the case of the conifold such a slicing is not convenient as one of the NS5's is parallel to the slice, and so instead we consider slices $v + w = c$ where each slice corresponds to a complex number c. In this case each slice is dual to an A_1 singularity resolved by a 2-sphere with area $|c|$.

In addition there may be D4-branes wrapping the circle at that v which become D3-branes. If v extremizes the superpotential then there may also have been D4-branes stretching between the two NS5-branes, which become D5-branes wrapping the S^2 with a half-unit of $U(1)$ gauge flux. There are two ways to stretch between the two NS5's, corresponding to the colors of the two simple components of the gauge group. In IIB these correspond to the two orientations of the D5 wrapping about the 2-sphere.

As a consistency check, we note that two D4-segments connecting the NS5-branes on both sides can combine to create a single D4-brane that wraps the circle. T-duality must commute with this process, and indeed it does. The two resulting D5-branes have opposite orientations and so annihilate. However the worldvolume $U(1)$ fluxes of $1/2$ and $-1/2$ mean that the unstable $D5 - \overline{D5}$ pair carries a single unit of D3-brane charge. And so in the end a D3-brane remains, which is the T-dual of the D4-brane wrapping x^6 as required.

Now it is easy to watch the cascade. As we probe lower energies the NS5-branes sweep out x^6, crossing each other. In the T-dual IIB this means that each step in the cascade corresponds to a change of the B-field on one 2-cycle by a single unit. At a slice corresponding to an extremum of the superpotential, such as $c = 0$ in the conifold configuration, there may be m D4-branes extending between the NS5's. As the superpotential is extremized the integral of the Kahler form over the 2-cycle vanishes. However the B-field and in particular

[9]If on the way from one NS5-brane to the other a D4 wraps the circle k times then there are k D4's that wrap the circle, and so the T-dual carries k units of D3-charge. This is why the D3-charge (and hence the rank of the gauge group) falls as the D4 unwinds.

the complex Kahler form $K + iB$ does not generally vanish and so we will say that the T-dual configuration is a D5-brane wrapped on this 2-cycle, although in light of the vanishing real area of the two-cycle one may consider this to be a fractional D3-brane. During each step of the cascade the NS5-branes separate and come together again on the other side of x^6, and so the distance increases from 0 to 1, which is identified with 0 because the x^6 direction is compactified. Correspondingly the pullback of the B-field to the worldvolume of the D5-brane decreases from 0 to -1, which may be identified with 0 via a large gauge transformation if at the same time we decrease the integral of the worldvolume gauge field F of each D5 by one unit. Note that such a shift is consistent with the shifted quantization condition (5.1). The D3-charge carried by each D5-brane is the integral of the F flux over the transverse 2-cycle and so the D3-charge decreases by one unit per D5-brane, that is the total D3-brane charge decreases by m units in keeping with the fact that there are m less D4's wrapping x^6. This yields a decrease in the rank of each simple gauge group by m units and so we have reproduced the cascade.

As noted above, instead of a time-dependent process this may be treated as a radius-dependent process. The integral of the B-field over the 2-sphere scales as the logarithm of the radius [33]. As we flow into the IR the 2-sphere wrapped by the D5-brane shrinks, at large N because the IR corresponds to smaller radii and at small N because in the T-dual MQCD the $|v|$ coordinate of the corresponding M5-brane shrinks, and so the B-field felt by its worldvolume shrinks as well.

In the radial picture the nonbaryonic vacua are easier to understand. Nonbaryonic vacua are T-dual to configurations in which k D4-branes end on two converging NS5-branes in the M-theory lift, like the m D4-branes at the $r = 0$ vacuum studied above. Thus when the NS5-branes converge the D4-branes are annihilated. In IIB this means that the corresponding k D5-branes do not fall to smaller radii like their compatriots that are connected across the NS5s. Instead these k D5-branes decouple at lower radii than the transition (lower than the corresponding condensate).

So what are the allowed values of k? To reproduce all of the vacua we need to T-dualize each segment of D4-brane separately, so that we find $m + n$ D5-branes and n anti D5-branes, each with half a unit of worldvolume field strength to reproduce the correct D3-brane charge

$$Q_{D3} = n \int_{S^2} (-F) + (n+m) \int_{S^2} F = \frac{-n}{2} + \frac{m+n}{2} = \frac{m}{2}. \tag{5.2}$$

Only the $n + m$ D5-branes are dual to the D4-branes that are shrinking to zero length, and so $0 \leq k \leq m + n$. Once the probes have become so small that the k D5-branes have been decoupled, $m+n-k$ D5-branes remain. If $k > m$ then $m+n-k < n$ and so there is more tension on the NS5-branes from the outside than the remaining tension on the inside, thus they change direction and the remaining $SU(m + n - k) \times SU(n)$ theory continues to cascade. This is the IR effective theory of the $r = m + n - k$ vacuum of the original $SU(m+n)$ with $2n$ flavors. The allowed range of k is $m \leq k \leq m+n$ and so $0 \leq r \leq n$.

If $k = m$ then $m + n - k = n$ and so once k D5-branes have decoupled we are left in a conformal theory which is a possibly strongly coupled $SU(n)$ gauge theory with $2n$

flavors. This is just the $r = m + n - k = n = N_f/2$ vacuum of the original $SU(m+n)$ theory, which is always a conformal field theory. Finally if $k < m$ then $m + n - k > n$ and so the NS5-branes cross as in the baryonic root vacua of a $SU(m+n-k) \times SU(n)$ theory. This leaves the Seiberg-dual $SU(n-m+k) \times SU(n)$, and so this case corresponds to the $r = n - m + k$ vacuum of the original theory. In all three cases we have found that the $SU(n+m)$ theory flows to an $SU(r)$ theory with $0 \leq r \leq n = N_f/2$ and N_f flavors, where the first and third cases contained some redundant descriptions of the same vacua. We have then reproduced the known set of vacua of an $N = 2$ supersymmetric $SU(n+m)$ gauge theory with $2n$ ($2n < n+m$) flavors when it is softly broken to $N = 1$ by an adjoint scalar mass.

5.2 The Cascade from Dielectric NS5-Branes

How is this description of the cascade related to the KPV dielectric NS5-brane description of Ref. [13]? In the current description the D5-brane wraps a cycle whose B-field changes by one unit. By Stokes' theorem we may equivalently say that the D5 sweeps out a single unit of H flux

$$\Delta \int_{S^2} B = \int_{S^2 \times I_{time}} H \tag{5.3}$$

destroying D3-charge because the D5 worldvolume coupling

$$S_{D5} \supset \int B \cup C_4 \tag{5.4}$$

implies that worldvolume B flux carries D3 charge. The origin of the H flux in M-theory was clear, it was dual to the bending of the M5-brane. However when we reduced to 10-dimensions it was obscured, as the bending was no longer visible in a single frame.

To see the origin of the H flux, as in M-theory, we must try to understand how the configuration relaxes. First we note that the tension of the D5-brane includes a contribution from its D3-brane charge, and so in particular is monotonically increasing in the B-field for positive $B + F$ and is not periodic in B. This is clear, for example, from the Dirac-Born-Infeld action on the D5-brane worldvolume. Thus any process that decreases the B-field on the D5-brane will reduce the energy. In the T-dual IIA description this matches our notion of relaxing, as decreasing the B-field corresponds to contracting the D4-branes, a process which is driven by the 4-brane tension. Thus we are searching for a process in IIB that creates H flux with 2-coordinates along the wrapped 2-cycle and also one coordinate along the time direction (or along the radial description in AdS/CFT-like descriptions of the RG flow as a flow to smaller radii).

Such a process is a Brown-Teitelboim [34] dielectric NS5-brane. This NS5-brane nucleates as a small, spherical NS5-brane that carries no net NS5-brane charge. In IIB NS5-branes carry $U(1)$ gauge fields and, as this NS5-brane was created from uncharged heat, the integral of this $U(1)$ gauge field over the 2-sphere must vanish. If the NS5-brane expands and then contracts out of existence such that its worldvolume links the stack of m D5-branes then the integral of G_3 over its worldvolume is m, where smaller cascades may result from NS5-branes that only wrap some of the m units of D5-brane charge. To find the $r = 0$ case it suffices to probe smaller distance scales than the radius of D5's before any NS5-branes have a chance to nucleate. Any NS5-branes nucleating at smaller scales will not link any D5-branes and so do not affect the gauge symmetry.

We will consider the maximal wrapping, again corresponding to vacua at the root of the baryonic branch of the $N = 2$ vacua. In this case the integral of C_2 over the wrapped sphere at any moment in time increases from 0 when the NS5-brane was first created to m at the end, at which point the volume of the S^2 shrinks to zero and so C_2 must blow up. However C_2 is not gauge invariant, rather $F + C_2$ is gauge invariant. This means that we may keep C_2 finite if we perform a gauge transformation during the NS5-brane's lifetime that decreases the integral of C_2 by m units but increases the integral of F by the same m units. The integral of F is the worldvolume D3-brane charge, and so when the NS5-brane disappears it carries m units of D3-brane charge. Of course this choice does just correspond to a gauge choice, which is consistent with the S-dual twisted K-theory fact that D3-brane charge in this background is only conserved modulo m. However the choice of a gauge in which C_2 does not blow up appears to coincide with a choice of gauge group such that the gauge coupling is finite.

This is not exactly the process that we are looking for. The problem is that the dielectric NS5-brane that wrapped the 3-cycle creates a single unit of H flux, that by a choice of boundary conditions we may consider to extend along the 2-cycle and time. Thus the integral of B over the two-cycle shifts by one unit for each such nucleation as we allow the system to time-evolve. This is similar to what we found above using T-duality, but the sign is wrong. The shift *increases* the D3-brane charge in the D5 stack by m units. The fact that the total D3-brane charge is conserved is S-dual to the following fact. On a $spin^c$ manifold, if the H flux is sourced by a stack of NS5-branes then a stack of Dp-branes that is nontrivial in homology but trivial in twisted K-theory may decay, but this decay leads to a worldvolume flux on the NS5-brane stack that carries precisely the lost charge, and so the Dp-brane charge is conserved whenever the H flux is sourced [8]. In our case the G_3 flux is sourced by D5-branes and so D3-brane charge is conserved when one counts the contribution from the MMS instanton NS5-brane and also from the D3-brane charge of the D5-branes.

To support the claim that the D3-brane charge is conserved we consider the case in which the D3-brane emitted from the NS5-brane is then absorbed by the stack of D5's. This is related by a Hanany-Witten transition to a configuration in which the D5-branes and NS5-brane do not link at all, in which case clearly no charges could have changed. In Fig. 5 we see that even the case in which the D3 is not reabsorbed is equivalent to one in which the stack of D3's escape from the D5's, a process which surely preserves the D3-charge. At finite g_s each D3-brane is then seen to be just the thin tube of dielectric D5 and NS5-brane [35] created when a D5 and NS5 appear to cross that ensures that in fact they do not cross. The existence of such tubes ensures that at finite g_s the linking of numbers of NS5-branes and D-branes never change, and so the non-linked situation is continuously related to our situation in which the branes appear to link but are actually unlinked by a stack of D3-branes.

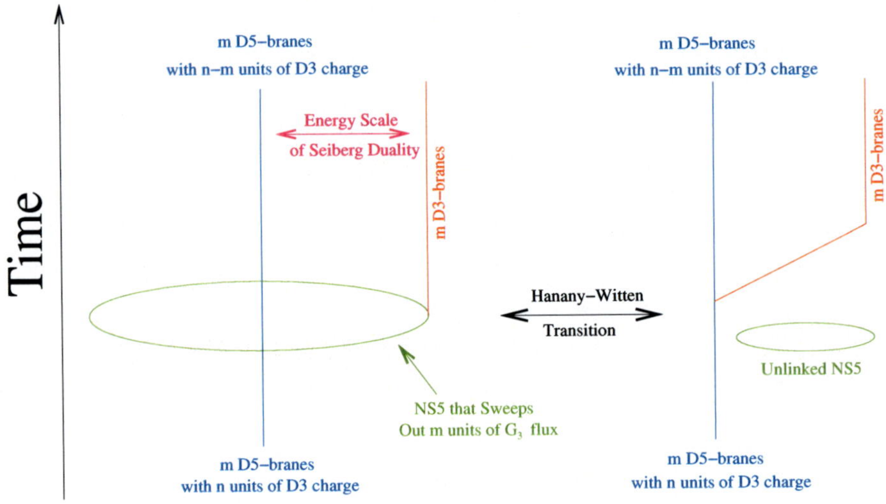

Figure 5: The cascade occurs when an NS5-brane bubble spontaneously forms in the vacuum and sweeps out a 3-surface linking the D5-branes. The NS5-brane wraps a trivial cycle and so carries no net NS5-brane charge, but the G_3 flux of the D5-branes produces D3-brane charge on the NS5-branes. Similarly the H flux of the NS5-branes destroys an equal amount of D3-charge on the D5-branes, reducing the gauge group as seen by a probe not big enough to extend to the NS5-branes from the D5-branes, where the gauge theory lives (the horizon at large m). The fact that the total D3-charge is conserved, as is clear to a UV observer who notices no reduction in gauge group, is apparent from the Hanany-Witten transition to a stack of D3-branes escaping.

The solution [13] is to consider an instantontic NS5-brane with the opposite orientation to that considered above. This NS5-brane will gain m units of D3-brane charge while the D5-branes will, as desired, lose m units of D3-brane charge. Again the D3-brane charge is conserved, and so one might think that the rank of the gauge group is not reduced. However the gauge theory lives on the D5 and D3-branes, and so is coupled even at low energies to the D3-branes created on the D5-brane worldvolume. However the IR degrees of freedom do not couple the gauge theory to the worldvolume $SU(m)$ gauge theory of the D3-branes contained in the NS5-brane, because the distance (in the w-plane, where the dual quarks live) to the NS5-brane is the length scale at which this step of the cascade occurred. Thus a probe at a higher energy than the energy at which this Seiberg duality occurred would couple to all of the D3-branes and thus not see a reduced rank, but this is what we expect as only probes at energy scales below that of the duality see the reduced gauge group.

The fact that the distance to the NS5-brane is the energy scale of the Seiberg duality is visible from the T-duality to MQCD. These NS5-branes are a remnant of the part of the M-theory configuration that we omitted when we reduced our configuration to type

IIA. This is particularly clear if we use the viewpoint in which the RG flows with the radius. Now the NS5-branes are static and the whole configuration is T-dual to the following improved IIA reduction of the M-theory configuration. Instead of making the NS5-brane continue, completely flat, forever along the v plane we instead approximate the M5 by a series of steps. In particular the flat sheet at fixed x^6 only continues up to the radius of the Seiberg-duality transition. At the boundary of this disc we may bend the NS5-brane at a 90-degree angle, and attach a cylindrical NS5-brane which wraps the x^6 circle once and then bends again to sweep out an annulus in the v plane, and then another cylinder, and so on approximating the M-theory curve. Each such cylinder contains m units of D4-brane charge, as it wraps the M-theory circle m times each time that one travels around the origin of the v plane once. As the cylinder is an NS5-brane that lies along the direction that is T-dualized, its T-dual will be another NS5-brane. The m units of D4-brane charge carried by the NS5-brane will dualize to m units of D3-brane charge carried by the NS5-brane in IIB. Intuitively the smearing of these D3-branes along the w-plane at the length scale of the Seiberg duality may be responsible for the mass gap of their worldvolume theory, this is in accord with the fact that the gauge symmetry is restored if the $N = 2$ supersymmetry is restored by projecting to $w = 0$.

The cylinder was always topologically trivial in IIA, that is it carried no net NS5-brane charge, and so the T-dual will also carry no NS5-brane charge. Thus we have found our stack of m D3-branes that have dielectrically expanded into an NS5-brane. As desired the distance to the branes on which the gauge theory lives is the scale of the Seiberg duality because the step was placed so as to approximate the M-theory curve.

At energy scales above that of the Seiberg duality, probes are larger than the sphere swept out by the NS5-brane. Thus the probes are larger than the distance between the D3-branes that have disappeared and those that have appeared, and so they do not register any loss of D3-brane charge. Correspondingly, at energy scales above the transition the rank of the gauge group is not reduced.

In more general cascades there will be several stacks of D5-branes wrapping the 2-spheres that resolve various singularities. These stacks will each carry some D3-brane charge, and if the D3-brane charge is high enough then NS5-branes will nucleate and then self-destruct, sweeping out a contractible 3-cycle that links some subset of the stack. Each linked D5-brane will lose a unit of D3-brane charge, and this will continue until such decay processes are no longer energetically favorable.

At large n the D5-branes are at the horizon. However the G_3 flux created by the D5-branes remains and the integral of the G_3 flux over the 3-cycle is still equal to m. Thus an NS5-brane sweeping out the 3-cycle still must emit m D3-branes before shrinking out of existence. Now the D3 charge of the D5's at the horizon drops and so the rank of each simple group has been reduced by m units. However at large n it is no longer clear how an NS5-brane may link less than the maximal m units of flux, and so the nonbaryonic roots are obscured, if indeed such vacua survive the large n limit.

6 Conclusion

We have argued that the WZW-like proposal for a continuous description of the cascade using the dielectric effect, in which the D3-branes embedded in the D5-branes decay via NS5-brane instantons, does not decrease the rank of the gauge group as desired. Instead it creates a flux that increases the rank of the gauge group in the IR, although at the original energy scale the gauge group is unaffected. Instead we have seen that the inverse process, in which dielectric NS5-branes nucleate and yield D3-branes, creates a flux that leads to less D3-brane charge embedded in the D5-branes where the gauge theory lives. Thus probes that are not sufficiently energetic to reach the NS5-branes will detect only the decreased D3-brane charge in the D5-branes, and will see the rank of the gauge group reduced as desired. The KPV proposal has the drawback that each step of the decay is a transition from a nonsupersymmetric configuration with anti-branes to a supersymmetric configuration, and so it may only proceed one step. In fact its range of validity, $m >> m - n$, ensures that it may only proceed for one step. Here no such restriction exists as the driving force for the process does not consist of anti-branes which are inserted by hand, but rather the tension of the M5-brane which is visible in the M-theory (and probably F-theory) lift but not in the IIA or IIB configuration. However by T-dualizing snapshots of the M-theory configuration reduced to IIA at different energy scales the effects of the truncated M-theory dynamics are retained.

Our argument is supported away from the large N limit, where traditionally such discussions have taken place, because it relies only upon the T-duality to MQCD. While the T-duality to MQCD has been heavily exploited in large N approaches [7, 30], it is unclear to what extent the MQCD approach is valid at large N. One obstruction is that the usual argument that allows one to pretend that spacetime is flat by introducing harmonic coordinates [36] needs to be consistent with the large N geometry. The large N limit has been critical to previous approaches, being used for example to justify the deformation of the conifold. However at radii smaller than the smallest $|v|$ of the curved M5-brane there can be no NS5-branes in the IIA reduction and so the T-dual does not contain a A_1 singularity. Thus one might suspect that the T-dual at such small radii, which correspond to energy scales below the last Seiberg duality, is a smooth space, and in particular that the conifold singularity is not there. Thus perhaps the deformation of the conifold is a straightforward consequence of the MQCD configuration. Of course such a minimum $|v|$ only exists if we turn on a superpotential, but this is when we expect the chiral symmetry breaking caused by this deformation.

In addition, as this argument is largely topological, it extends to $N = 2$ theories broken to $N = 1$ by an adjoint chiral multiplet mass μ. The original case is recovered in the limit of large μ. These configurations, in turn, are local descriptions of more general theories in which the $N = 2$ supersymmetry is broken to $N = 1$ by a superpotential which is polynomial in the adjoint chiral multiplets and sufficiently nondegenerate. Thus the T-duality described above may be applied to a neighborhood of the extrema of any such superpotential, the only places where nonconformal products of two simple gauge groups with bifundamental hypermultiplets may be found. The algorithm generalizing this approach to more than two gauge groups at the same extremum has been described above.

We have also seen that in the MQCD picture, in addition to the baryonic root vacua (with the $r = 0$ vacuum chosen at the last step) that define the Klebanov-Strassler cascade, there are many other cascades corresponding to different vacua. In particular we have described the simplest cascade,

$$SU(n) \times SU(n+m) \longrightarrow SU(n) \tag{6.1}$$

which corresponds to the $r = 0$ nonbaryonic root of the $SU(n+m)$. Combining this with the baryonic root description, we can conclude for example that the Klebanov-Strassler cascade can stop at any step if the $r = 0$ vacuum is chosen. It may be interesting to use the MQCD picture to classify all of the cascades and to compare them with the field theory, where one has a choice of vacua after each duality.

An independent application is to the twisted K-theory classification of D-branes, where it is a first example of three phenomena. First, this example is evidence for an S-duality covariant generalization of the K-theory classification, and in fact is inconsistent with the usual K-theory classification. Second, the fluxes produced by the MMS instanton, which are sometimes dismissed as irrelevant, here are seen to play a key role. The lost D3-brane charge on the D5-stack in fact is not destroyed directly by the MMS instanton, but on the contrary is destroyed by the flux created by the instanton. In fact the cancellation between the charge-violation of an MMS instanton and the charge violation on the worldvolume of the source of the flux it wrapped plays the physical role of ensuring that in the UV the original gauge group is restored. That is, no D3-branes are really ever destroyed, it is just that our probes get shorter and so we can not reach them all anymore as we flow into the IR.

Finally in the large N near-horizon case we see the (S-dual) K-theory classification at work, classifying fluxes. However the branes classified by the (S-dual) K-theory are not in the spacetime at all, but rather live at the horizon. Previously K-theory classifications of D-branes were restricted to finite stacks that live inside the spacetime, in fact in the approach of [37] it is crucial that the branes are inside of some sphere.

The two big questions are left untouched. First, what is the S-duality covariant K-theory classification? Second, what is the F-theory lift of the conifold configuration? It is easy enough to fiber the torus over 9 of the dimensions, this just gives the M-theory configuration. But is its fibering over the IIB circle even well-defined? The IIB circle consists of the set of maps from the F-theory torus to the E_8 fibered over M-theory, so a point in the total space consists of a 9-dimensional point plus a point in the E_8 fiber and so a section of an E_8 bundle over the 9d space with the F-theory branes describing the obstructions to creating this bundle from the 11d bundle. But when is such a construction possible?

Acknowledgements

I would like to thank Nick Halmagyi for posing this problem, and Sunil Mukhi for telling me that KPV solved it first. I would also like to thank M. Douglas and I. Klebanov for discussing this proposal. This work has been partially supported by IISN - Belgium (convention 4.4505.86), by a "Pôle d'Attraction Universitaire" and by the European Commission RTN program HPRN-CT-00131, in which I am associated to K. U. Leuven.

References

[1] R. Minasian and G. Moore, *K-theory and Ramond-Ramond charge*, *J. High Energy Phys.* **11** (1997) 002, [arXiv:hep-th/9710230].

[2] E. Witten, *D-Branes and K-Theory*, *J. High Energy Phys.* **12** (1998) 019, [arXiv:hep-th/9810188].

[3] J. J. Manjarin, *Topics on D-brane Charges with B-fields*, [arXiv:hep-th/0405074].

[4] E. Witten, *Baryons and Branes in Anti de Sitter Space*, [arXiv:hep-th/9805112].

[5] J. Maldacena, G. Moore and N. Seiberg, *D-Brane Instantons and K-Theory Charges*, [arXiv:hep-th/0108100].

[6] J. Evslin, *Twisted K-Theory from Monodromies*, [arXiv:hep-th/0302081].

[7] I. R. Klebanov and M. J. Strassler, *Supergravity and a Confining Gauge Theory: Duality Cascades and χSB-Resolution of Naked Singularities*, [arXiv:hep-th/0007191].

[8] J. Evslin, *IIB Soliton Spectra with All Fluxes Actived*, [arXiv:hep-th/0211172].

[9] C. Vafa, *Evidence for F-Theory*, [arXiv:hep-th/9602022].

[10] E. Diaconescu, G. Moore and E. Witten, *E_8 Gauge Theory, and a Derivation of K-Theory from M-Theory*, [arXiv:hep-th/0005090].

[11] J. Evslin and U. Varadarajan, *K-Theory and S-Duality: Starting over from Square 3*, *J. High Energy Phys.* **03** (2003) 026, [arXiv:hep-th/0112084].

[12] O. Loaiza-Brito, *Instantonic Branes, Atiyah-Hirzebruch Spectral Sequence, and SL(2,Z) Duality of N=4 SYM*, [arXiv:hep-th/0311028].

[13] S. Kachru, J. Pearson and H. Verlinde, *Brane/Flux Annihilation and the String Dual of a Non-Supersymmetric Field Theory*, [arXiv:hep-th/0112197].

[14] P. Bouwknegt, A. Carey, V. Mathai, M. Murray and D. Stevenson, *Twisted K-theory and K-theory of bundle gerbes*, *Comm. Math. Phys.* **228** (2002) 17-45, [arXiv:hep-th/0106194].

[15] V. Mathai and D. Stevenson, *On a Generalized Connes-Hochschild-Kostant-Rosenberg Theorem*, [arXiv:hep-th/0404329].

[16] D. Marolf, *Chern-Simons Terms and the Three Notions of Charge*, [arXiv:hep-th/0006117].

[17] W. Taylor *D-Branes in B Fields*, [arXiv:hep-th/0004141].

[18] A. Kapustin, *D-Branes in a Topologically Nontrivial B-Field*, [arXiv:hep-th/9909089].

[19] D. S. Freed and E. Witten, *Anomalies in String Theory with D-Branes*, [arXiv:hep-th/9907189].

[20] A. Sen, *Tachyon Condensation on the Brane Antibrane System*, [arXiv:hep-th/9805170].

[21] P. Hořava, *Type IIA D-Branes, K-Theory and Matrix Theory*, [arXiv:hep-th/9812135].

[22] J. M. Maldacena and C. Nunez, *Towards the large N limit of pure N=1 super Yang Mills*, [arXiv:hep-th/0008001].

[23] A. Hanany and E. Witten, *Type IIB Superstrings, BPS Monopoles and Three-Dimensional Gauge Dynamics*, [arXiv:hep-th/9611230].

[24] A. Yu and V. Schomerus, *D-Branes in the WZW Model*, [arXiv:hep-th/9812193].

[25] C. Bachas, M. Douglas and C. Schweigert, *Flux Stabilization of D-Branes*, [arXiv:hep-th/0003037].

[26] E. Witten, *Solutions of Four-Dimensional Field Theories Via M-Theory*, [arXiv:hep-th/9703166].

[27] J. Evslin, H. Murayama, U. Varadarajan and J. .E. Wang, *Dial M for Flavor Symmetry Breaking*, [arXiv:hep-th/0107072].

[28] N. Seiberg, *Electric-Magnetic Duality in Supersymmetric Non-Abelian Gauge Theories*, [arXiv:hep-th/9411149].

[29] G. Carlino, K. Konishi and H. Murayama, *Dynamics of Supersymmetric $SU(n_c)$ and $USp(2n_c)$ Gauge Theories*, [arXiv:hep-th/0001036].

[30] K. Oh and R. Tatar, *Duality and Confinement in N=1 Supersymmetric Theories from Geometric Transitions*, [arXiv:hep-th/0112040].

[31] M. Atiyah and E. Witten, *M-Theory Dynamics on a Manifold of G_2 Holonomy*, [arXiv:hep-th/0107177].

[32] P. C. Argyres and M. R. Douglas, *New Phenomena in SU(3) Supersymmetric Gauge Theory*, [arXiv:hep-th/9505062].

[33] I. R. Klebanov and N. Nekrasov, *Gravity Duals of Fractional Branes and Logarithmic RG Flow*, [arXiv:hep-th/9911096].

[34] J. D. Brown and C. Teitelboim, *Neutralization of the Cosmological Constant by Membrane Creation*, *Nucl. Phys.* **B297**, 787, (1988).

[35] N. R. Constable, R. C. Myers and O. Tafjord, *The Noncommutative Bion Core*, [arXiv:hep-th/9911136].

[36] G. W. Gibbons, *Branes as BIons*, [arXiv:hep-th/9803203].

[37] G. Moore and E. Witten, *Self duality, Ramond-Ramond fields, and K-theory*, *J. High Energy Phys.* **05** (2000) 032, [arXiv:hep-th/9912279].

INDEX